WHAT'S NEXT?

Max Brockman is a literary agent at Brockman, Inc., which represents Jared Diamond, Nassim Taleb, Richard Dawkins, and Steven Pinker, among others. He also works with the Edge Foundation, Inc., a nonprofit foundation that publishes the Edge newsletter (http://www.edge.org). A graduate of the University of Pennsylvania in 2002, he lives in New York City.

WHAT'S NEXT?

WHAT'S NEXT?

Dispatches on the Future of Science

EDITED BY MAX BROCKMAN

Vintage Books

A Division of Random House, Inc.

New York

To my parents

CONTENTS

PREFACE

Most of us have learned from long experience that in all walks of life it's important to engage with the thinking of the next generation, to better understand not just what is going on in our own time but what issues society will face in the future. This exercise is especially valuable in science, where so many of the important discoveries are made by those in emerging generations. To see what today's brilliant young scientists are up to is the impetus for bringing together this collection of brief essays.

The eighteen young scientists featured here are investigating a variety of questions that will have long-term and fundamental effects on the way we live—and even on how we see ourselves and our place in the universe. Their ideas will eventually help to redefine who and what we are.

To generate this list of contributors, I approached some of today's leading scientists and asked them to name some of the rising stars in their respective disciplines: those who, in their research, are tackling some of science's toughest questions and raising new ones. The list that resulted amounts to a representative who's who of the coming generation of scientists.

I have asked each of the contributors to write about the questions they're asking themselves with respect to their field of study. Their essays are especially fresh, as many of these authors have not yet had the time or opportunity to write for a broad nonacademic audience. Among others, you will hear from:

- David M. Eagleman, director of the Laboratory for Perception and Action at Baylor College of Medicine, analyzing how time is perceived by the brain;
- Katerina Harvati, a paleoanthropologist at the Max Planck Institute for Evolutionary Anthropology, examining the evidence of past hominid extinctions and what it means for the future of our own species;
- Matthew D. Lieberman, associate professor of psychology at UCLA, wondering if the physical structure of the brain makes us more likely to form and cling to certain ideas;
- Sean Carroll, a senior research associate in physics at Caltech, discussing what we still don't know about the origins of our universe—and about the arrow of time;
- Laurence C. Smith, professor and vice chairman of geography and professor of earth and space sciences at UCLA, hypothesizing that climate change could cause a land rush into the newly economically viable Northern Rim;
- Lera Boroditsky, an assistant professor of psychology, neuroscience, and symbolic systems at Stanford University,

looking at how the languages we speak shape the way we think; and

• Sam Cooke, a postdoctoral associate in neuroscience at MIT, speculating on when and how we will be able to manipulate our own memories—and whether or not we should.

The enthusiasm of these young researchers for their work, and their passion for science, is evident. Their bold exploration of new ideas and their attempts to stretch the boundary of what is known are inspiring. It is my hope that *What's Next?* will give readers a head start on comprehending what may well be in store for all of us in the future.

Max Brockman
New York
January 2009

WHAT'S NEXT?

LAURENCE C. SMITH

earned a Ph.D. in earth and atmospheric sciences at Cornell University in 1996 and is now professor and vice chairman of geography and professor of earth and space sciences at UCLA. He has published more than fifty research papers in journals such as Science *and* Nature. *In 2006 he briefed Congress on the likely impacts of northern climate change, and in 2007 his work appeared prominently in the Fourth Assessment Report of the United Nations' Intergovernmental Panel on Climate Change (IPCC).*

Smith's work on northern climate change has been funded by the Guggenheim Foundation, the National Science Foundation, and the National Aeronautics and Space Administration. Honors include a NASA Young Investigator Award (2000), finalist for NASA's Presidential Early Career Award (2002), and a Bellagio residency from the Rockefeller Foundation (2007).

WILL WE DECAMP FOR THE NORTHERN RIM?

LAURENCE C. SMITH

Like so many other cultural shifts, it gathered long, then broke quickly. At last the world—including a majority of people in the United States—has acknowledged that global warming is real.

Changing the public's opinion was not easy. It took the work of thousands of scientists, painstakingly accumulated over more than three decades. Their findings were then steadily communicated to the world through massive synthesis reports in 1990, 1995, 2001, and 2007 by the Intergovernmental Panel on Climate Change (IPCC), displaying a level of organization unprecedented in science. These reports document the evidence, now overwhelming, of our new man-made climate.

Pivotal to the public opinion shift were ardent "Third Culture" scientists—among them James Hansen, at NASA's Goddard Institute for Space Studies; Lonnie Thompson, at Ohio State University; Richard Alley, at Pennsylvania State University; and Mark Serreze, at the University of Colorado—with a

talent for grasping the most significant discoveries and chan-
neling them to the public through books, interviews, YouTube,
and popular magazines like *Rolling Stone*. These efforts at
public outreach represented a significant shift in the culture
of science. As a graduate student in the mid-1990s, I witnessed
the widespread, if subtle, scorn directed at the remarkable
astronomer and writer Carl Sagan by his professional colleagues
for his efforts in publicizing his scientific work. But today, and
especially in climate-change science, public outreach is part of
the job and a cause for appreciation and emulation by scientific
colleagues.

Other events, largely unforeseen, also figured prominently
in converting the public. The graphic horrors wrought by Ka-
trina—regardless of that hurricane's cause—sowed national
unease via millions of televisions and computer screens. The
failed presidential bid of Al Gore in 2000 freed him to film *An
Inconvenient Truth* in 2006—and, together with the IPCC, he
won the 2007 Nobel Peace Prize. The 2006 decision of Wal-
Mart to embrace and aggressively market green technology
reached millions more, including many who wouldn't be caught
dead at an Al Gore movie. In my home state of California,
Republican governor Arnold Schwarzenegger asserted, "The
[climate] debate is over"—and from a scientific and public-
opinion standpoint, he was right.

The burden of proof is past, so what's next? The debate has,
if anything, intensified; the line of scrimmage has simply
moved downfield. Questions like "Is it real?" and "Is it our
fault?" have morphed into "What will happen?" "Where?"
"How fast?" and "What are we going to do about it?" Science
may have led us to these questions, but our answers will rever-

berate far beyond science. At stake is no less than the global pattern of human settlement in the twenty-first century.

So, what *will* happen? Here is what we know currently: First, the warming is just revving up. It is 90 percent certain that continued greenhouse gas emissions at or above the current rates will induce far greater climate change in the twenty-first century than we've yet experienced.[1] In every plausible population-growth or greenhouse-gas-emission scenario for the next century (barring some as-yet-undiscovered nonlinearity in the climate system), basic physics dictates that Earth's climate must continue to warm, with global average temperatures rising between 1.8°C and 4.0°C (3.2°F and 7.0°F) by the end of this century.[2] How high we go depends on how much carbon we choose to load into the atmosphere; the lower value is the IPCC's optimistic estimate, which assumes a stabilized global population and the adoption of clean-energy technology. The high value is the estimate based on unabated dependence on fossil fuels.

If those temperature changes don't sound large to you, they should. Even the most optimistic number (1.8 °C) triples the warming we had in the twentieth century. Furthermore, thanks to the long life of greenhouse gases in the atmosphere and the sluggish response of the world's oceans, we are already "locked in" to much of that warming, regardless of what policy changes we enact. The global temperature increase to 2030 is all but

[1] S. Solomon et al., eds., "Summary for Policymakers," in *Climate Change 2007: The Physical Science Basis Working Group I Contribution to the Fourth Assessment Report of the Intergovernmental Panel on Climate Change* (New York: Cambridge University Press, 2007), 13.
[2] Ibid. "Technical Summary," 70.

committed, and even if we could cap greenhouse gas emissions immediately at 2000 levels, we would still experience about half the projected warming by midcentury. But over the long run, policy changes will have a large impact: only 20 percent of the projected 2100 temperature rise is currently locked in. At this point, it is still possible—through aggressive societal action—to blunt the warming. But we cannot stop it.

The hotter temperatures will increase evaporation, drying soils and raising the frequency of drought, especially in two broad belts from 20° to 40° north and south latitudes—that is, in both hemispheres. The number of extremely dry days will increase sharply in the southwestern United States, southern and eastern Europe, southern Africa, and eastern South America.[3] Water vapor in the air will also increase, in obedience to the Clausius-Clapeyron equation, which states that the water-holding capacity of the atmosphere must go up 7 percent for every 1°C rise. Because water vapor fuels weather systems, the frequency of extreme precipitation events—and therefore floods—will go up right along with it. Deadly, power-sucking heat waves—like the killers in France in 2003, the United States in 2006, and Japan in 2007—will happen more often. Sea level will continue to rise (it's rising now, around three millimeters per year), the only uncertainty being exactly how fast and how high. Low-elevation coastal areas, including Florida, the Netherlands, island nations, and impoverished Bangladesh, will face inundation in the coming decades.

If you saw *An Inconvenient Truth* or read climate-change stories in the press, you already know most of this bad news.

---

[3] Solomon et al., fig. 10:18.

Alongside more speculative notions about hurricanes and wild-fires, they are the most widely reported scientific predictions for the twenty-first century. However, even these are not the starkest forecasts of our climate models. The most robust changes will sweep across the northern high latitudes, starting at about the 45th parallel running through the northern United States, Canada, Russia, and Europe. North of that line, the climate changes will be unrivaled on Earth. Temperatures will rise at nearly *double* the global average—driven mostly by milder winters—and precipitation will increase sharply as well.

Already the impacts are obvious in the extreme north, where melting Arctic sea ice, drowning polar bears, and forlorn Inuit hunters are by now iconic images of global warming. The rapidity and severity of Arctic warming is truly dramatic. However, the Arctic, a relatively small, thinly populated region, will always be marginal in terms of its raw social and economic impact on the rest of us. The greater story lies to the south, penetrating deeply into the "Northern Rim," a vast zone of economically significant territory and adjacent ocean owned by the United States, Canada, Denmark, Iceland, Sweden, Norway, Finland, and Russia. As in the Arctic, climate change there has already begun. This zone—which constitutes almost 30 percent of the Earth's land area and is home to its largest remaining forests, its greatest untouched mineral, water, and energy reserves, and a (growing) population of almost 100 million people—will undergo one of the most profound biophysical and social expansions of this century.

There are a number of reasons why these northern latitudes have never been a magnet to southern settlers. Sunlight is strongly seasonal. In the extreme north and within deep continental interiors, permafrost (permanently frozen ground) ratchets up construction costs and keeps soils waterlogged, making the land a moist heaven to billions of mosquitoes. Growing seasons are short, agricultural yields are low, and large tracts of land are mountainous. But the single greatest inhibitor to southern forms of life—plant, animal, and human—is the mind-numbing, crushing cold of winter. Summers are warm, even hot (there are air conditioners throughout the Northern Rim), but winters are a frost-panting monster. Deciduous trees crackle and die, frogs freeze solid in their mud beds, and at −40° (the Celsius and Fahrenheit scales converge at this number), compressors fail, steel shatters, and manual work becomes impossible.[4]

"Minus-forty" is feared and hated by everyone who has experienced it. The shutdown of human activity it mandates has been described to me by people from all around the Northern Rim—restaurateurs in Whitehorse, Cree trappers in Alberta, truck drivers in Russia, retirees in Helsinki. And while they express otherwise mixed opinions on the various problems and opportunities presented by recent warming, the one sentiment they all share is utter relief that minus-forties are becoming increasingly rare.

By defanging winter's bite, will global warming spawn new human societies around the Northern Rim? The idea is not so far-fetched: According to an August 8, 2007, article in the *Wall Street Journal*, speculative real estate buying in Newfoundland and Labrador rose sharply in 2007. Likewise, the *Financial Times* of December 1, 2007, reports rising real estate markets across

---

[4] F. Hill and C. Gaddy, *The Siberian Curse* (Washington, DC: Brookings Institution Press, 2003), 41–49.

northern Norway, Sweden, and Russia. But before you rush to Realtor.com to scope out acreages near Anchorage or Winnipeg, listen up: yes, there will be growth, but it won't happen everywhere. Like human expansion over the millennia, its direction will be shaped by the choices we make and the previous imprint of history and geography.

In his book *Collapse*, my UCLA colleague Jared Diamond scours human history to identify five prime factors that determine the likelihood that an existing society will fail: environmental damage, loss of trade partners, hostile neighbors, climate change, and how a society chooses to respond to its environmental problems. Any of these, alone or in combination, can trigger a society's collapse. Turning the question around, what makes a new society likely to successfully establish itself? First and foremost is economic opportunity, followed by environmental suitability, opportunities for investment and trade (implicit in this is military security and the consistent rule of law, without which investors balk and trade will not be stable), friendly neighbors, and willing settlers.

At present, these requirements are met only to varying degrees around the Northern Rim. Abundant economic opportunities exist in the form of commodities fossil fuels, minerals, fish, and timber—and, indeed, their exploitation currently generates most of the Northern Rim's gross domestic product, the second contributor being government services. The neighbors are generally friendly; relative to the rest of the world, all eight Northern Rim nations have low internal unrest and share amicable borders—though Finland frets over its long border with Russia, and Russia worries about the United States and (especially) China on its thinly populated eastern flanks.

Nonetheless, there have been no serious military incursions among the eight countries since World War II. Seven of them (Russia is the exception) enjoy the most stable political systems and rule of law in the world.

That leaves environmental suitability, trade, and settlers. It seems likely that climate warming will blunt the most significant environmental limitation—brutal winters—currently restricting human expansion in the Northern Rim. The warming will also ameliorate other problems, such as the short growing season, and create new ones, such as pest infestations, but these are secondary to the impact of milder winters. So climate change, one of five key factors known to collapse past societies, will actually engender them in these northern latitudes.

Still remaining are trade and settlers. All else in place, these factors depend mainly on markets, infrastructure, and demographic trends. While commodity prices are famously volatile, over the long haul (this century and the next, for example), it's a safe bet that the world's demand for water, minerals, energy, food, and timber will remain high. But demand per se does not create trade; there must also be infrastructure, without which commodities cannot get to markets, and settlers, without which there is no labor. And sufficient settlement requires domestic population growth, immigration, or both. Since strong contrasts in both infrastructure and demographic trends exist around the Northern Rim today, I expect them to shape the geographical pattern of northern human expansion. Unless, that is, we jolt the system somehow—which is exactly what we did twice in the last century, which led to the geographical contrasts in the first place.

The Northern Rim has long been resource rich and

population poor, an irresistible siren call to its central governments. Over the years, their efforts to increase infrastructure, population, and economies have been motivated by different ideologies and have had mixed results. Also, their treatment of aboriginals has varied hugely, with U.S. and Canadian groups faring best in recent years, followed by those in Scandinavia and lastly Russia. But all this pales in comparison with two enormous choices made in the twentieth century—choices that utterly transformed the footprint of humanity in the Northern Rim. They were the U.S. Army's decision to occupy Canada during World War II and Joseph Stalin's decision to create the Gulag, a string of forced-labor camps and exile towns across Siberia, between 1929 and 1953.

The underlying motivation for Stalin's murderous prison-camp system ran far deeper than the silencing of political malcontents. It was nothing less than a forced settlement of his country's remote territories—then sparsely occupied by aboriginals—with ethnic Russians. The Gulag was responsible for some of the worst atrocities of modern history, including countless deaths from starvation, exposure, exhaustion, and outright murder. But as a forced-settlement tool, the program was a resounding success. By the early 1950s, the camp population stood at 2.5 million people, most of them political exiles or people condemned for minor crimes.[5] They labored in mines, cut timber, and built roads, railroads, and factories. If they survived their sentences, ex-prisoners were legally prohibited from returning home. The towns grew huge, and by the end of the 1980s they were major cities, entrenched across some of the

[5] Hill and Gaddy, 86.

coldest terrain on Earth: Novosibirsk, Omsk, Yekaterinburg, Khabarovsk, Chelyabinsk, Krasnoyarsk, Noril'sk, Vorkuta. Mother Russia had urbanized Siberia.

Today the future of these cities lies in doubt. Their locations are arbitrary, selected more for quaint socialist ideals, such as taming nature and Engel's dictum (the idea that industry should be uniformly distributed across a country), than for the pragmatic requirements of economic viability. They exist in places that don't make sense: in harsh environments, at long distances from one another and from trading partners, precariously linked by absurdly stretched infrastructure that requires deep subsidization from Moscow. The burden socialist planners placed on the Soviet economy by founding these cities in such inhospitable locations was so great that in their book *The Siberian Curse*, Fiona Hill and Clifford Gaddy argue that "the cost of the cold" deeply saddled the Soviet economy and helped bring down the USSR in 1991.

After the Soviet collapse, the subsidies disappeared. Throughout the 1990s, the giant Siberian cities emptied out faster than Detroit in a bad layoff year. Today there are some signs of population stabilization, and limited prosperity is trickling back with high oil prices. A second, wiser attempt at massive infrastructure establishment in Siberia may well occur in this century, as Russia and China eye its vast natural wealth. There are already early indications of this: Vladimir Putin has officially opened a 6,200-mile highway between Moscow and Vladivostok—the longest highway in the world—and Russian military scholars have floated the idea of a "free economic zone" in Siberia's Far East, opening up the region's vast timber reserves to development with Chinese capital. But the population continues to drop, among ethnic Russians and aboriginals alike—a decrease compounded by high mortality, suicides, and grinding poverty. Any new human expansion in Siberia will require a considerable installation of infrastructure, a wiser settlement plan, abandonment or relocation of failed towns, and reversal of today's negative population trends.

In North America there was a different story. The small number of white settlers initially coexisted with aboriginal groups, grubbing out a frontier living in mining, trapping, fishing, and small farming. World War II changed everything. At the height of the war, the U.S. Army injected massive infrastructure into the Northern Rim—airstrips, roads, bases, pipelines, ports, radar stations, and towns throughout Alaska, Canada, Greenland, and Iceland. There were sixty thousand U.S. military personnel and contractors in Alaska alone. In northwest Canada, there was what amounted to a friendly occupation by forty thousand soldiers and civilian workers—a huge population infusion for such a sparsely populated area. The Canadian government watched from Ottawa as the United States transformed its country with the Northwest Staging Route, the Alaska Highway, the Canol Pipeline, and dozens of other projects.[6] The imprint of the U.S. Army Corps of Engineers still shapes the pattern of Northern Rim human settlement and economic activity today.

Aboriginal groups, utterly marginalized, saw their homelands overrun. In Canada they endured culturally damaging resettlement programs. However, since the early 1970s they

[6] See K. S. Coates and W. R. Morrison, *The Alaska Highway in World War II: The U.S. Army of Occupation in Canada's Northwest* (Toronto: University of Toronto Press, 1992).

have steadily regained control of their territories and mineral rights through legal action (land claims), now largely concluding. U.S. and Canadian aboriginals are forming blue-suit business corporations and strengthening their economic and political clout, and they have some of the highest population growth rates of the Northern Rim. If the American droughts continue, I and my fellow climate-change refugees will be unable to count on swarming across our northern border: all the good spots will be taken.

So what does this tell us about the coming northern expansion? The United States and Canada have sensible infrastructure, working rules of law, and a fast-growing domestic population of suitable settlers. These two countries are well positioned for expansion. Russia, with badly placed infrastructure and a severely declining population, is not. Scandinavia, with its well-developed roads, ports, and universities, is already poised to benefit from the winter climate warming already in the pipeline.

Lest you misinterpret this essay, let me make something absolutely clear: the transformation I describe is a conversion from land that is hardly livable to land that is somewhat livable. There will be no northern utopia. Global society will be far better off if we can stay where we are, enjoying the much larger and friendlier landmass between about 50° N and 45° S. The 23.5° axial tilt of our planet dictates that there will always be darkness and cold at higher latitudes, even if greenhouse warming causes February in Churchill, Manitoba, to warm up to, say, February in Minneapolis. And the die is not fully cast: yes, we are locked into a large amount of warming, but not (as of yet) so much that we must all migrate to Yakutsk. Just as Stalin shaped the Northern Rim with prison camps and the U.S. Army shaped it with airstrips and the Alaska Highway, our choice is whether to shape it with sensible, far-reaching plans based on the IPCC's optimistic or pessimistic greenhouse-gas-emission scenario. At best, the Northern Rim will become tolerable, not a paradise. I do not advise buying acreage in Labrador.

But maybe in Michigan.

CHRISTIAN KEYSERS,

*born in 1973 in Belgium, studied psychology and biology in
Germany and in Boston. He obtained a Ph.D. in neuroscience
at the University of St. Andrews, Scotland, in 2000 and then
worked in Parma, Italy, where he contributed to the discovery
of auditory mirror neurons and enlarged the concept of mirror
neurons by applying it to emotions and sensations. He is the scientific director of the Neuroimaging Center of the University
Medical Center Groningen, the Netherlands, where he is full
professor of the social brain. He has received the Marie Curie
Excellence Award of the European Commission and is associate editor of the journal* Social Neuroscience.

MIRROR NEURONS

Are We Ethical by Nature?

CHRISTIAN KEYSERS

Shared Circuits: How You Invade My Brain

In the 1990s, a pivotal discovery was made by the Italian neuroscientists Giacomo Rizzolatti, Vittorio Gallese, Leonardo Fogassi, and their colleagues at the University of Parma, where they were investigating how the brain controls our actions. Using very thin electrodes, they measured the activity of single neurons in a region of the monkey brain called the premotor cortex. This region contains neurons that are active while the monkey grasps or manipulates objects. Some of these neurons, for instance, are active when the monkey grasps a peanut, others while the monkey breaks the peanut's shell. It is the activity of these neurons that triggers the monkey's movements. Humans also have a premotor cortex, and if a surgeon stimulates this region while a patient is undergoing surgery, the patient reports feeling the urge to perform certain actions. The premotor cortex is a key element of our voluntary actions and of our personal control over our own bodies—a stronghold of our free will.

The surprise came when one of the experimenters grasped a peanut to give it to the monkey. The very same neuron that had responded when the monkey grasped a peanut also responded when the monkey simply saw someone else perform the same action. At first, Rizzolatti et al. did not believe their own finding. How could a brain region involved in voluntary actions respond while someone else's actions were simply watched? What finally convinced them was a remarkable congruence: neurons involved in performing a particular action (e.g., grasping) would respond to the sight only of that particular action. This could not be coincidence.

The experimenters later showed that even the *sound* of someone else's actions would activate neurons involved in executing those particular actions, and it became increasingly clear that the monkey's brain could transform the actions of the experimenters into motor programs that the monkey would use to perform the same actions.

We call such neurons *mirror neurons*, because through them the motor activity in the brain of the monkey mirrors the actions of others.

A number of experiments have shown that humans have a similar system. Motor representations of our own actions are activated whenever we see the actions of other people, animals, or even robots.

If motor representations of our actions are activated while viewing those of others, why don't we always overtly imitate other people? The answer is that while we're observing the actions of others, a neural gate seems to block the output of our motor areas, keeping our bodies from imitating the movements we are seeing. Behind this gate, our brain can covertly share the actions of the people around us. We no longer only see the movements of others, we also feel their movements inside us, as if we were doing the same thing they were. If we are in the starting blocks of a race, for instance, the false start of a competitor automatically almost triggers our own. When we see people dance, we often cannot help but feel our own bodies moving. Our motor system is thought of as the seat of our free will, but each time I witness your movements, you permeate this stronghold. Your acts become mine, and my acts become yours.

This phenomenon is not restricted to physical movement. When someone taps my shoulder, say, my somatosensory cortex makes me feel the sensation. But simply seeing someone else being tapped activates the same area of my brain. If I cut my finger, my cingulate cortex and anterior insula will register the pain—and these areas also become active if I see you cut your finger. The vicarious representations are not quite as strong as those produced when we experience our own sensations, but we nevertheless feel a milder version of what the other person feels. I feel a sensation something like pain in my finger when I see you cut yours. My own shoulder itches as I watch a tarantula crawl across James Bond's shoulder in *Dr. No*.

Moreover, our emotions seem to obey a similar rule. When I smell a foul odor, for example, my insula produces a feeling of disgust. The very same region is active when I see an expression of disgust on your face; it is as if I were experiencing your disgust. These shared neural activations go hand in hand with our individual subjective experience: who has never felt his or her mood improve at the sound of someone else laughing, or felt sadness when a friend cries? The emotions of others are contagious because our brain activates our own emotions at the sight of them.

These brain circuits can keep us from seeing other individuals as something "out there." Indeed, we are able to feel their actions, sensations, and emotions inside us, as if we were in their shoes. Others have become us.

Shared Circuits Allow Us to Learn from Others

What are these shared circuits good for? Humans owe their success to their ability to cooperate with and learn from one another. Hunting is a good example: given spears and coordination, humans can bring down buffalo or woolly mammoths. While the media often glorify the genius of individuals and Nobel Prizes are given to the inventors of new ideas, most useful things (such as spears) are the result of thousands of years of slow technological improvement—of learning from a more experienced teacher and then adding one's own innovation and teaching it to the next generation.

Somehow our brains must be able to learn from others. This process is far from trivial. To learn to make a spear, for instance, we have to convert the sight of someone else's manipulations into something very different—the nerve impulses required to move our own hands in similar fashion. Mirror neurons appear to have solved this difficult task: each time we see an action, they transform this sight into the motor commands necessary to replicate the action. While we watch an expert perform a series of action components, one after another, until a spear is made, our brain activates similar action components in the same order, composing the novel sequence of spear-making from the familiar movements of picking up a stone, sharpening it, picking up a stick, and tying the stone to the stick.

The sharing of others' emotions is an important element in this process. Virtually all animals learn based on trial and error. When we share the actions and emotions of others, this ancient mechanism becomes social learning. If we see someone taste an unfamiliar fruit and look pleased, our brain will share the action and the positive consequence as if we ourselves had eaten the fruit and enjoyed it. If we see the person expressing disgust, we will share this negative experience. Our vicarious trial-and-error learning mechanisms tell us that eating that particular fruit is either a good or a bad idea, providing us with all the benefits of learning without the risk of being poisoned.

These brain circuits thus blur the bright line between your experiences and mine. Our experiences fuse into the joint pool of knowledge that we call culture. With the advent of language, books, and television, this sharing becomes global, allowing us to exchange experience across time and space.

Shared Circuits Create an Ethical Instinct

Often we are more than passive observers; many of our decisions affect other people. Imagine I am one of two starving individuals and I find a piece of food. Shall I eat it all, or shall I share it?

From an individualistic point of view, eating it all seems the

most rational decision. It saves me from starvation, and the
only disadvantage is that the other person might die—but that
is not my problem, is it? If I share the food, I may hope that the
other person will feel obliged to return the favor later on—but
who knows, he might not.

Because of these shared circuits, though, the equation is
different. Since the same brain areas are active whether we are
feeling our own pain or witnessing that of others, this means
that the vicarious sharing of others' feelings is not an abstract
consideration but a toned-down equivalent of our own. If I eat
all the food, I will not only witness but also share my companion's suffering, whereas if I divide the food I will share his joy
and thankfulness. My decision is no longer guided only by my
hunger but also by the real pain and pleasure my companion's
pain and pleasure will give me.

What's more, although the strength of these brain mechanisms varies from person to person and also depends on the circumstances, all individuals show some degree of sharing. It is
thus reasonable to assume that other people will also take my
suffering into account while deciding whether or not to share
food with me. Indeed, experiments performed in a wide variety
of societies show that, in most places in the world, people will
tend to share the wealth with others instead of keeping it all for
themselves.

The circuits allowing us to share knowledge with others
therefore have another profound implication for human
nature: they lay the foundation for an *intuitive altruism*. Most
cultures have what is called an ethical Golden Rule. Christianity, for instance, urges, "Therefore all things whatsoever ye
would that men should do to you, do ye even so to them: for this

is the law and the prophets" (Matthew 7:12), while Islam holds, "Not one of you truly believes until he wishes for others what he wishes for himself" (Muhammad, 13th Hadith of Nawawi). I believe that the brain mechanisms that make us share the pain and joy of others are the neural bases that intuitively predispose us according to this maxim. Our brain is ethical by design.

This is not to say that humans are incapable of hurting one another. Indeed, if our personal interest is in direct conflict with that of others, our desire for benefits might outweigh our empathy. Unfortunately, human ingenuity has designed methods to reduce empathy in those circumstances where it fails to serve our purpose. In the military, the distance that separates the generals from the human suffering their armies cause minimizes their empathy and favors self-interested decisions. At the same time, the chain of command strips moral responsibility from the soldiers who do directly witness the suffering. In such a way, empathy can be bypassed in the service of efficiency. The development of weapons that kill at a distance has a similar effect. Insights into the biology of our empathy help us realize the risk of such distancing and point us toward ways to build the natural mechanisms of empathy into our institutions.

Humans are the result of evolution, and evolution favors individuals who will leave more offspring, not altruistic individuals who forget their own interests in the service of others. At first, a brain that forces us to share the emotions of others seems at odds with survival of the fittest. Humans, however, are social animals. Imagine a family whose members can share the actions, sensations, and emotions of other family members. And imagine a family without this ability. In the latter, brothers will steal from one another, cheat one another, and learn very little from one another. The family with strong sharing will respect one another's needs, learn from and collaborate with one another. In the face of hardship, this capacity for cooperation will prove essential for hunting, gathering, and child care—and this family will therefore leave more offspring. As individuals, we may pay the price of shared misery because of our ability to empathize, but we reap all the benefits that our culture and a stable society bestow. Mirror neurons—and their gift of insight into the emotions of others—enable us to manipulate other individuals but also prompt us to use this understanding for good and not for evil. It is heartening to realize that there is something inside us that makes us truly care about other individuals as if they were part of our extended self.

NICK BOSTROM

is a philosopher and the director of the Future of Humanity Institute at Oxford University; before moving to Oxford, he taught philosophy at Yale University. He received his Ph.D. from the London School of Economics in 2000. He has a background in physics, computational neuroscience, and mathematical logic, and his research covers issues in the foundations of probability theory, global catastrophic risk, the ethics of human enhancement, and the effects of future technologies.

Bostrom is coeditor (with Julian Savulescu) of Human Enhancement *and (with Milan Cirkovic) of* Global Catastrophic Risks, *and he is the author of several influential papers on human enhancement; his writings have been translated into sixteen languages. He has also worked briefly as a consultant for various organizations, including the European Commission and the U.S. Central Intelligence Agency.*

HOW TO ENHANCE HUMAN BEINGS

NICK BOSTROM

The Wisdom of Nature

Medical science is difficult. We know this because, despite our best efforts, it often fails. Yet medicine typically aims merely to fix something that's broken. Human enhancement, by contrast, aims to take a system that's not broken and make it better—in many ways a more ambitious goal.

Human beings are a marvel of evolved complexity. Although great strides have been made in the biological sciences, most of this complexity still remains to be unraveled. When we manipulate complex, evolved, poorly understood systems, our interventions often fail or backfire. Pain medications turn out to cause birth defects or high blood pressure; allegedly safe pesticides are discovered to be carcinogenic; infant formula is invented and touted as superior to mother's milk, yet later found to be lacking in some fatty acids that are important for brain development.

The list of such failures is long. Iatrogenesis, adverse effects from medical treatment, has been estimated to cause some 225,000 deaths annually in the United States, making it the third leading cause of death. It seems as if there is a "wisdom of nature" that we ignore or override at our peril. The belief in nature's wisdom—and corresponding doubts about the prudence of tampering with nature, especially human nature—often manifests as diffusely moral objections to enhancement. Many people have intuitions about the superiority of "the natural" and the troublesomeness of human hubris. Some might base these ideas on theological doctrine, but often there is no such underpinning; often there is nothing more than a discomfort with altering the status quo.

Enhancement enthusiasts—optimistic about the promise of biomedical intervention to improve cognition, enrich emotional well-being, or retard aging—may be tempted to dismiss as mere superstition these preferences for the "natural," or to resignedly conclude that the difference in outlook rests on a basic moral disagreement about which no reasoned discussion is possible. But we can do better than that. We can recognize that there is a grain of truth in the idea that nature has some wisdom. By better understanding the extent and limitations of this wisdom, we can develop a practical heuristic, or rule of thumb, for identifying promising human enhancements. This heuristic is not a substitute for ordinary forms of medical research—clinical trials and the like—but it can guide research by identifying the parameters within which interventions have the greatest chance of working and highlighting the potential side effects and risks associated with various types of intervention, so that we can proceed with extra caution whenever these warning signs are present.

By considering our species' origins and the shortcomings of the evolutionary process that brought us to where we are, we can more easily spot cases where nature's way may not have delivered a wise solution to a particular problem. In such cases—and only in such cases—developing workable human enhancements might be relatively easy.

Evolution as a Great Engineer

Evolution can be likened to a surpassingly skillful engineer. The limitations of this metaphor are part of what makes it useful for our purposes.

Evolution led to a system, the human organism, which is far more complex than anything we humans have built to date. We marvel at the complexity of the human organism, how its various parts have evolved to solve intricate problems: the eye to collect and preprocess visual information, the immune system to fight infection and cancer, the lungs to oxygenate the blood. The human brain—the focus of many of the most alluring proposed enhancements—is arguably the most complex object in the known universe.

Given our rudimentary understanding of the human organism, particularly the brain, how can we hope to enhance such a system? It would amount to outdoing evolution, and we might well doubt our ability to do that with our present tools and level of scientific understanding. I suggest we turn this vague scruple into a question, which we should ask about any proposed enhancement intervention. We can call it the *evolutionary-optimality challenge*: if the proposed intervention would result in an enhancement, why have we not already evolved to be that way?

There are three categories of potential answers:

- changed trade-offs
- value discordance
- evolutionary restrictions

If any of these categories offer an explanation for why a desired enhancement has not already evolved, then that enhancement intervention can be considered promising. It still needs to be tested in clinical trials, but the heuristic gives us a green light to proceed to that stage. By contrast, proposed enhancements for which no such answer is found are likely to fail: they might not work at all, or they might have serious side effects (which could be subtle or take years to manifest). In such cases, the heuristic warns us to proceed with extra caution or to try something else.

Changed Trade-offs

The human organism evolved to operate in a particular environment: a hunter-gatherer life on the African savannah. Now it must function in the modern world, a very different environment. Modern conditions are too recent for our species to have fully adapted to them; thus the trade-offs that evolution struck may no longer be optimal. We may be able to make some tweaks and adjustments that would fit the human organism better to its new environment, even though our engineering talent falls far short of that of the evolutionary processes that created the original design.

One novel feature of modern life for most people in developed countries is the abundant availability of food independent of season. In the state of nature, by contrast, food is relatively scarce much of the time, making energy conservation paramount and forcing difficult energy-expenditure trade-offs between metabolically costly tissues, processes, and behaviors. For example, the human brain constitutes only 2 percent of body mass yet accounts for over 20 percent of total energy expenditure. (In newborns, brain metabolism accounts for a staggering 60 percent of total metabolism.) The brain, heart, gastrointestinal tract, kidneys, and liver together consume 70 percent of basal metabolism. Evolution had to make a difficult trade-off between the size and capacity of these organs and the allocation of time and energy to activities other than searching for food and maximizing its nutritive value.

Suppose we consider a potential enhancement that promises to give us more mental energy. Apply this heuristic: "Why, if this is such a good thing, do we not naturally have a higher level of mental energy?" A plausible answer can be given involving changed evolutionary trade-offs. Greater mental energy, while perhaps in itself a benefit, would have come at the cost of increased metabolic expenditure (the brain consumes more energy during sustained mental activity). Today, of course, we no longer care about conserving calories (if anything, the opposite is true). It might therefore be an excellent bargain if we could find some stimulant that increased mental energy and burned more calories. Here the increased availability of food provides a new resource that changes the optimal point of that trade-off. It is not surprising that we already have stimulants such as caffeine and wakefulness enhancers such as modafinil, both of which seem to be effective at increasing mental energy with relatively few side effects.

Or it might be possible to design genetic enhancements of intelligence that come at the expense of increased head size or an extended period of maturation. We can easily see why we wouldn't have evolved those traits even though they would have been effective in increasing intelligence. While greater cognitive ability might have been advantageous to hunter-gatherers, the risks and burdens associated with a larger head or a longer period of immaturity would have outweighed the gains. With new resources, such as obstetric assistance, cesarean section, and safe environments for children, those burdens would now be far less significant.

Another potential enhancement would be to increase DNA repair activity in our cells. This could help prevent cancer and age-related degeneration. If the cost were merely to increase our required calorie intake, it would be a price well worth paying.

In addition to providing new resources, the environment can also change by placing new demands on us. Before the invention of written language, there was, of course, no direct selection for literacy. Numeracy, analytical skills, abstract thinking, and the ability to concentrate on a cognitive challenge for a long time are also much more useful in the contemporary world than they were in the Pleistocene. These changes in demand, too, suggest possible enhancements. For example, we might find a drug that improves our concentration. Why do we not already have a greater ability to concentrate, given concentration's apparent usefulness? Well, maybe there is a trade-off involved. Perhaps intense concentration increases brain metabolism, or maybe it reduces peripheral awareness. Calories are no longer scarce, and while peripheral awareness is important to hunter-gatherers (that lion is sneaking up behind you), in modern society it's more important to sustain concentration—on a textbook, say, or a computer terminal or an interlocutor. Again unsurprisingly, some people benefit from such concentration enhancers as nicotine or Ritalin. These drugs meet the evolutionary-optimality challenge.

Value Discordance

The second category of answers to the evolutionary-optimality challenge resides in the discrepancy between the standards we may wish to apply and the standards governing evolution. Evolution selects for inclusive fitness (that is, the fitness of the individual plus the fitness of her social partners weighted by their degree of genetic relatedness), whereas we are interested in optimizing for such values as well-being, achievement, knowledge, meaningful relationships, and moral excellence— these being far more important goals to us than bearing the largest possible number of offspring.

This discordance of objectives is a rich source of promising enhancements. In cases where we can point to such value discordance, we can meet the evolutionary-optimality challenge without necessarily supposing that our engineering talents exceed those of evolution. A mediocre technician who could never design a car, let alone a good one, may well be able to convert the latest BMW model into, say, a crude rain-collecting device. If we happen to value a rain-collecting device more than we value a car, this would count as an enhancement, from our point of view. Likewise, we might be able modify the human organism to better serve our idiosyncratic objectives even if we were unable to improve its performance as a survival-and-reproduction machine.

This won't work for all traits we might wish to enhance. Many traits we value would also have served to increase reproductive success in the early human environment. Health is one example. But many other traits we value would, on balance, not have contributed to evolutionary success; an obvious example is contraceptive technology. Vasectomy, birth-control pills, and other contraceptive methods can be thought of as enhancements that increase our control over our reproductive systems. We may value such enhancements because they facilitate family planning and increase choice. And it is no mystery why evolution has not already provided us with an easy reproductive off switch.

We can distinguish at least two distinct forms of value discordance. The first is that the characteristics that would maximize an individual's inclusive fitness are not always characteristics that would be best for her. The second is that they are not always those that would be best for her society, or impersonally best. Both forms of value discordance suggest interventions that might be workable enhancements from some point of view.

For example, we might devise interventions that would raise our boredom threshold. Such an intervention might make our lives more pleasant: we might find more things interesting and take more pleasure in work, hobbies, other people. Plausibly, a low boredom threshold was adaptive in the early human environment; it would have kept us from squandering time and energy on activities not materially contributing to survival and reproduction. The claim here is not that we should completely eliminate our capacity for boredom but that, on average, our boredom threshold is lower than what would make our lives most flourishing and fulfilling—and also that it might be feasible to raise it, perhaps by means of some pharmacological or genetic intervention.

We can also think of potential enhancements that might benefit society or humankind as a whole but where it is clear why evolution has not already provided us with them. For example, extended altruism and the ability to control violent impulses come in degrees. We can readily see how evolution would select for a degree of these traits that would maximize an individual's inclusive fitness. We can also see why society today would benefit if individuals had, on average, more extended altruism and were, on average, better able to control violent impulses. It might be both feasible and socially desirable to enhance these traits—for example, by means of empathogenic drugs, spiritual practice, improved upbringing, or cultural and social environments that strengthen these prosocial tendencies.

Even when a trait we wish to enhance would in itself have promoted inclusive fitness, we can sometimes still invoke value discordance to meet the evolutionary-optimality challenge. It might be that the trait is intrinsically coupled with another trait that would have been fitness-negative. Evolution would have struck one trade-off between these two traits, but if our priorities differ from those of evolution, we may have reason to strike a different trade-off. For example, suppose that unusually high levels of intellectual creativity come at the expense of a degree of absentmindedness. While we might, other things being equal, prefer not to be absentminded, we might be willing to accept a considerable degree of absentmindedness if we would thereby gain great intellectual creativity. If it could be shown that the preferred trade-off would have been maladaptive in the early human environment, it would give us grounds for being optimistic that we will find an intervention that gives us what we want. By contrast, we should be skeptical of a proposed intervention that seems to give us something for nothing—unless an answer to the evolutionary-optimality challenge can be found in one of the other two categories our heuristic provides.

Evolutionary Restrictions

The final category of potential answers to the evolutionary-optimality challenge is a little more theoretically complicated than the other two. It arises from the fact that evolution is not a perfect fitness-optimizing process. Evolutionary optimization is subject to a set of important limitations, and in some special cases these suggest that we can beat evolution at its own game. To continue the automotive analogy: we may be able to produce a better car even though our engineering skills are inferior to those of the team that originally built one—if we have access to better tools and materials.

Evolutionary restrictions fall into three subcategories:

- fundamental inability (evolution is fundamentally incapable of producing a particular trait);
- entrapment (evolution is stuck in a "local optimum" that excludes the trait); and
- evolutionary lag (the evolution of the trait takes so many generations that there has not yet been enough time for it

to develop and spread throughout the entire human
population).

First, consider fundamental inability. Biology is limited in
what it can build. For instance, it is unlikely that any terrestrial
organism could biologically produce a diamond or large metallic
objects; hence evolution cannot achieve diamond tooth enamel
or a titanium skeleton. If these enhancements became techni-
cally feasible, they would easily meet the evolutionary-optimality
challenge. Examples multiply. Evolution can probably not produce high-performance silicon chips to augment neural computation, even though such chips provide important benefits. While
such enhancements pose many technical challenges—how to
ensure that implants are biocompatible, for instance—there is
no special evolutionary reason, no wisdom-of-nature reason, for
thinking that such enhancements are not feasible.

Entrapment in a local optimum occurs because natural
selection is a myopic search process that can get stuck in a state
in which any small change would make the solution worse even
if some big change might make it better. One example is the
human appendix, a vestigial remnant of the much larger cecum
found in our herbivorous primate ancestors. Whereas the
appendix may have some limited immunological function, it
can easily become infected. In the state of nature, appendicitis
is a life-threatening condition and is especially likely to occur at
a young age. Shedding the appendix would likely have increased
fitness in the early human environment. However, a smaller
appendix *increases the risk* of appendicitis; carriers of genes predisposing for small appendices are at greater risk for appendicitis than noncarriers and thus, presumably, have lower fitness.

Unless evolution could find a way of eliminating the appendix in one fell swoop, it would be unable to get rid of the organ; thus it remains. An intervention that safely and conveniently removed the appendix might be an enhancement, increasing both fitness and quality of life.

Another way in which evolution can get locked into a sub-optimal state is through the phenomenon of heterozygote advantage. This refers to the common situation in which an individual who is heterozygous for a particular gene (that is, who carries two different versions, or alleles, of it) has an advantage over a homozygote individual (who has two identical copies). Heterozygote advantage is responsible for many cases of potentially harmful alleles being maintained at a finite frequency in a population. The classic example is the incompletely recessive sickle-cell allele: homozygote individuals suffer from sickle-cell anemia, a potentially fatal blood disease, while heterozygote individuals benefit, in this case from improved malaria resistance. Sickle-cell heterozygotes have greater fitness than both types of homozygote (those lacking the sickle-cell allele and those having two copies of it). Balancing selection preserves the sickle-cell allele in populations at a frequency that coincides geographically with the prevalence of malaria. The "optimum" that evolution selects is one in which, by chance, some individuals will be born homozygous for the allele, resulting in sickle-cell anemia. The "ideal optimum"—everybody being heterozygous for the allele—is unattainable by natural selection because of the Mendelian inheritance pattern, which ensures that each child born to heterozygote parents has a 25 percent chance of being born homozygous for the sickle-cell allele.

Heterozygote advantage suggests an obvious enhancement opportunity. If possible, the variant allele could be removed and its gene product administered as medication. Alternatively, genetic screening during in vitro fertilization could be used to guarantee heterozygosity, enabling us to reach the ideal optimum that eluded natural selection.

Then there is evolutionary lag, the final restriction of evolution's optimizing ability. Evolutionary adaptation takes time—often a very long time. If conditions change rapidly, the genome will fall behind. Given that conditions for anthropoid ancestors were highly variable because of migration into new regions, climate change, social dynamics, advances in tool use, and adaptations in pathogens, parasites, predators, and prey, our species has never been perfectly adapted to its environment. Evolution runs up fitness slopes, but with the fitness landscape continually changing, it may never reach a peak. Even if beneficial alleles or allele combinations exist, they may not have had time to diffuse across all human populations.

If we find a desirable gene that evolved recently and has not yet diffused through the entire human population, interventions that insert it into the genome or mimic its effects would likely meet the evolutionary-optimality challenge. A simple example is lactose tolerance. While development of lactose intolerance was adaptive for mammals as a way of promoting weaning, dairy products have stimulated selection for lactase in humans over the last five thousand to ten thousand years—not long enough for the trait to diffuse to all human populations. Taking lactase pills enables lactose-intolerant people to digest lactose, widening the range of food they can enjoy. This enhancement clearly meets the evolutionary-optimality challenge. Perhaps we can find similar cases: genes involved in brain development have also been under strong positive selection, with new variants emerging over the last thirty-seven thousand years (and possibly even more recently), a relatively short period in evolutionary terms.

To conclude: evolution's shortcomings ought to be viewed as our opportunities. By systematically considering the limitations of the evolutionary process that created the human organism, we can identify promising possibilities for enhancing it, using interventions that are feasible today or may become feasible in the relatively near future.[1]

In the longer run, we should be able to go far beyond these first steps. And once we obtain a relatively complete understanding of the human organism—or, alternatively, we learn how to build entirely artificial systems of equal complexity and performance—we would no longer need to lean on the evolution heuristic and "the wisdom of nature" for support. At some stage, we may learn how to design new organs and bodies ab initio. Someday we may no longer even rely on biological material to implement our bodies and minds. Freed from most practical limitations, the task would then become to make wise use of our powers to self-modify. In other words, the challenge would shift from being primarily scientific to being primarily moral. If that moral task seems comparatively trivial from our current vantage point, this might reflect our present immaturity.

---

[1] The ideas summarized here were developed in collaboration with Dr. Anders Sandberg. We also wish to acknowledge valuable input from Dr. Rebecca Roache and from participants at several conferences where the original research paper was presented.

SEAN CARROLL,

a senior research associate at Caltech since 2006, received his Ph.D. in physics in 1993 from Harvard University, with a thesis titled "Cosmological Consequences of Topological and Geometric Phenomena in Field Theories." He was a postdoctoral researcher at the Center for Theoretical Physics at MIT and at the Institute for Theoretical Physics at the University of California, Santa Barbara, and an assistant professor of physics at the University of Chicago. His research ranges over a number of topics in theoretical physics, including cosmology, field theory, particle physics, and gravitation. He is currently studying the nature of dark matter and dark energy; connections between cosmology, quantum gravity, and statistical mechanics; and the question of whether the early universe underwent a period of inflation.

Carroll has written a graduate textbook, Spacetime and Geometry: An Introduction to General Relativity, *and recorded a course on dark matter and dark energy for The Teaching Company. He has received fellowships from the Alfred P. Sloan Foundation and the David and Lucile Packard Foundation. Among his honors are the MIT Graduate Student Council Teaching Award and the Villanova University Arts and Sciences Alumni Medallion. In 2007 he was named NSF Distinguished Lecturer by the National Science Foundation. He is cofounder of and contributor to the Cosmic Variance blog http://blogs.discovermagazine.com/cosmicvariance/.*

OUR PLACE IN AN UNNATURAL UNIVERSE

SEAN CARROLL

The universe is trying to tell us something. But so far we're having trouble making out just what it's trying to say.

Science is about understanding, but progress in science is driven by misunderstanding. When observation after observation agrees with our predictions, it's hard to move forward, but when an experimental result flies in the face of our favorite theories, we begin getting somewhere. No matter how good the theories are, they rarely encompass every phenomenon we might hope to explain; our goal is to expand the reach of our theories into the unknown, and nothing is more helpful in that quest than a stubborn fact that refuses to fit the current picture.

Sometimes, however, we're not so lucky. We have theories that, within their zone of applicability, do a good job of explaining what we observe but leave us with a nagging feeling that they're not the final story. A perfect example is the standard model of particle physics. Put together in the 1960s and '70s, it has heroically managed to account for mountains of data accumulated by particle physicists in the 1980s and '90s. Only recently—with the discovery of neutrino masses and evidence of the existence of dark matter and dark energy—has the standard model begun to stumble, and we are still far from knowing what new theory might replace it.

But nobody ever really thought the standard model was the last word in particle physics anyway. For one thing, it doesn't include a fundamental force—gravity—so it can't be considered complete. For another, it doesn't seem completely "natural." It contains a lot of particles, described by a lot of parameters, and there doesn't seem to be much rhyme or reason in the values of those parameters. There are a variety of symmetries—some broken, some unbroken—but no compelling reason why. There is one crucial scale of energy for the weak nuclear force, and this scale is enormously smaller than you might expect if you hadn't measured it.

Even though the standard model has passed a withering battery of experimental tests, it just doesn't smell right. There must be some simpler and more robust arrangement underlying this rambling chaos.

Naturalness is in the eye of the beholder; what looks perfectly natural to one person might seem deeply suspicious to another. But physicists have pretty clear criteria for deciding whether something is natural or not. Any good physical theory comes with a set of numbers—the *constants of nature*, at least within the purview of that specific model. If one or more of those numbers are extraordinarily large or small, we begin to think the model is not natural. In particular, we wonder whether nature has given us a clue to some piece of physics underlying what we see—one that explains those provocatively large or small numbers in simple terms.

A striking example of an unnatural number is provided by *vacuum energy*—the energy density of empty space. We're used to thinking of energy as associated with some sort of stuff—particles or radiation or motion. But in the context of general relativity, Einstein's theory of space and time, energy is inherent in the fabric of space itself. How would we know? According to Einstein, every form of energy contributes to the expansion of space. Vacuum energy—which remains constant in density, whereas ordinary matter and radiation disperse—imparts a persistent impulse to the expansion of the universe, causing distant galaxies to accelerate away from us. This acceleration of the universal expansion was actually detected in 1998, a demonstration that the hypothetical concept of vacuum energy is strikingly real.

But it's not natural. The problem is not with the idea of vacuum energy itself but with the amount of vacuum energy that accounts for the acceleration: it's much smaller than it should be. We can assemble the measured values of other constants of nature—the speed of light, Newton's gravitational constant, Planck's constant from quantum mechanics—to predict the value that the vacuum energy should have. It works out to be about 10^{112} ergs per cubic centimeter. If that seems like a big number, it is; the *measured* value is only about 10^{-8} ergs per cubic centimeter. That's a difference of a factor of 10^{120} (10 followed by 120 zeros) between two numbers that should be about the same. What's going on?

Perhaps the universe is telling us something. The observed

vacuum energy seems unnaturally small, but there's a hitch: if it were much bigger, we wouldn't be here to observe it. If the vacuum energy were anywhere close to its "natural" value of 10^{112} ergs per cubic centimeter, space would accelerate at such a fantastic rate that individual atoms would be ripped apart, not to mention planets, stars, and galaxies.

So maybe we just got lucky. Or maybe we didn't get lucky at all—perhaps we observe an unusually small value for the vacuum energy because we live in an unusual part of the universe.

The observable universe—the part we see around us—seems remarkably similar over very large scales. There are about the same number of galaxies nearby as there are in regions of space billions of light-years away. But we don't see the whole universe; light travels at a finite speed, so there's a limit to how far our observations can reach. Beyond that observable horizon, the universe might well stretch out indefinitely and might not be uniform at all. In fact, the local laws of physics and constants of nature might change from place to place. Imagine a patchy universe—one in which conditions and parameters are uniform within individual patches but the patches themselves are all completely different. We would, of necessity, find ourselves only in a hospitable patch. Perhaps the value of the vacuum energy appears unnatural to us because we are witnessing a dramatically unrepresentative bit of an overwhelming ensemble, the whole of which is as natural as anyone could want.

Or . . . perhaps not. This kind of anthropic reasoning— attempting to explain unusual features of the observable universe in terms of selection effects applied to a larger, unobservable "multiverse"—leaves a lot of people feeling queasy, if not downright hostile. For one thing, invoking a multitude of regions with different local laws of physics in order to explain a handful of numbers doesn't seem very economical. For another, it would be very hard ever to know whether such reasoning was on the right track. The problem with invoking a multiverse in order to explain why conditions allow for our existence is that whether or not there *is* such a multiverse, we still wouldn't be here speculating about such things if the constants of nature weren't consistent with our existence. An ensemble of patches of space with different local laws of physics is one explanation for unnatural parameters; another is that the laws of physics are unique — and just *happen* to be compatible with the existence of intelligent life. How are we ever to choose between these two explanations?

A possible clue is provided by another unusual feature of our observable universe: the big bang. Cosmologists use the phrase in two separate senses. The big bang itself is that purported singular moment at which the universe came into being, while the big bang model is the scenario describing the evolution of our universe from an initially hot, dense state to the cool, expanding conditions we see today. We have very little idea what happened at the very beginning of the universe's history, so the moment of the bang is purely conjectural, but the evolution of the universe from very early times to today is well understood. The big bang model is on firm empirical ground, even if the bang itself is simply a placeholder for our ignorance. Nevertheless, just as with the standard model of particle physics, the big bang model's success at fitting the data doesn't mean that it doesn't present thorny issues of naturalness.

One of the underpublicized puzzles of modern cosmology is that we don't know *why* the early universe looked like it did.

Here we're not talking about constants of nature, as we were in the case of vacuum energy, but about the configuration in which the universe finds itself. Right now, that configuration looks like a thinly spread collection of gas and stars and dark matter (a form of matter that has not yet been directly observed but whose gravitational effects on galaxy structure have been), collected into galaxies scattered throughout the universe, evolving in a background of vacuum energy. The whole shebang is expanding and cooling off; therefore things were hotter and denser in the past. They were also smoother; the universe started out almost completely uniform, and the inexorable pull of gravity, over the fourteen billion years since the bang, gradually assembled stars and galaxies. As far as we can tell, the universe will continue to expand forever. Galaxies will move farther and farther apart, and eventually the universe will become completely smooth again, as all matter dissipates into the void. In the far distant future, the universe will be a thin gruel of elementary particles, growing ever colder and more distant from one another.

One feature of this story is so obvious that it needs to be pointed out for you to take any notice of it: the past is very different from the future. The early universe is hot and dense; the late universe is cold and dilute. Well . . . why is it like that? The truth is, we have no idea.

The distinction between past and future is so deeply ingrained in the way we think about the universe that it doesn't seem to need explaining. Like fish oblivious to the water, we are hardly aware of the deep puzzle of temporal asymmetry—the so-called *arrow of time*. But it is the most salient feature of our local physical environment. We can turn eggs into an omelet, but we can't turn the omelet back into eggs. We can remember the past but not the future. As anyone who has seen a movie run backward knows, time cannot simply be reversed without quickly running into absurdities. There is no symmetry between past and future in the dynamics of our local environment. What is puzzling is that there *is* such a symmetry in the fundamental laws of physics; they work equally well running forward or backward in time. If the laws of physics tell us that past and future are created equal, why does everything about our everyday experience tell us something different?

The origin of the arrow of time, right up to the moment when we crack an egg for breakfast in the morning, can be traced back to the beginning of the universe. Physicists keep track of the progress of time in terms of *entropy*—a measure of the disorder of some physical situation. If you scatter some collection of objects, they will wind up in any number of disordered arrangements—arrangements to which we assign a high degree of entropy. But there are only a few ways to order the objects precisely, and the entropy of such orderly arrangements is correspondingly low. An ordered deck of cards has low entropy, while a well-shuffled deck has high entropy. Higher-entropy configurations, if you will, are more natural than lower-entropy ones, for the simple reason that there are a lot more such high-entropy configurations to be found.

Because high-entropy configurations are more natural and more numerous, the entropy of an isolated physical system tends to increase (or at least not decrease) over time. That's the second law of thermodynamics, perhaps the most cherished principle in all of physics. The second law is what guarantees that omelets don't spontaneously turn into eggs; eggs are much lower in entropy than omelets, since there are far fewer ways for the constituent molecules to arrange themselves into an egg than into an omelet.

But that's only half the story. Although entropy tends to increase because there are more ways to be high-entropy than low-entropy, that doesn't explain *why entropy was low in the first place*. But it was. The entropy of our universe started out preposterously small and has been growing ever since. Think about it: our observable universe contains a huge number of particles (about 10^{88}, to be precise) spread out over billions of light-years. Yet at early times they were all delicately squeezed into a hot, dense plasma in an extremely small region of space. How unnatural is that? There are many more ways for all those particles to be spread far apart from one another, and that's exactly what's happening as time goes on. Our present universe is somewhat medium in entropy, compared with what it could be. In the future, entropy will be enormous, as it evidently prefers to be, but in the past it was remarkably small.

So not only are some of the constants of nature apparently fine-tuned, but so was the state of the universe at early times. *Why hasn't the universe been in a high-entropy state all along?* There's no obvious reason why the universe's matter and radiation ever had to be squeezed together; it could just as well have been thinly spread out for all eternity. Then there would be no arrow of time at all; the universe would simply sit there, with no distinction between past and future, and nothing would ever happen.

You can guess where this is going. In a truly high-entropy universe, where nothing ever happens, one of the things that wouldn't happen is life. Every feature we think of as characterizing life—metabolism, reproduction, evolution, memory—depends deeply on the arrow of time. Life, it is no exaggeration to say, is propelled through time by the growth of entropy. If entropy were large all along, it wouldn't grow and there would be no life. Can we use anthropic reasoning to explain the low entropy near the big bang?

Not exactly. Despite the worry that anthropic reasoning is inherently untestable, it does come with a certain logic: if something unnatural is necessary for the existence of life, we should observe it to be sufficiently fine-tuned to account for a hospitable universe but no more so. That's precisely what we do find with the observed value for the vacuum energy: if it were larger by any appreciable amount, we wouldn't be here to talk about it.

But the case of entropy is different. The early universe is not only in a very special configuration, but its entropy is far lower than it needs to be to account for our existence. At the most, our existence requires the kind of environment we find in the Milky Way galaxy. But that's only one of 100 billion galaxies in our observable universe; what are they all for? In particular, why is the matter collected into these galaxies not spread uniformly and thinly throughout space, as it would be in a truly high-entropy universe? Even if life requires a universe with a low-entropy beginning, our actual universe seems needlessly profligate in the amount of fine-tuning at early times.

We've now moved into waters where we not only lack well-supported theories but don't even have well-articulated speculations. Be that as it may, more and more cosmologists are taking seriously the question of what happened *before* the big bang. Remember that the bang itself is by no means well understood;

we tend to think of it as a moment of infinite density and curvature, but the truth is, we just don't know. We don't understand how to reconcile the demands of quantum mechanics with the curved space-time of general relativity; once we do, what seems singular and forbidding might be resolved into something smooth and continuous. Increasingly, physicists are daring to stretch their imaginations beyond the veil of the bang to suggest that our observable universe had an unobserved prehistory.

Within that prehistory, we might find an explanation for the fine-tuned, low-entropy state of our early universe. Imagine a universe that was in a purportedly high-entropy state: cold, dilute, with all particles widely separate from one another. But remember, there is still vacuum energy, so even empty space is not perfectly quiescent. The rules of quantum mechanics tell us that in the presence of such energy, there are irreducible fluctuations in the vacuum. Particles pop into and out of existence, and fields occasionally arrange themselves in statistically unlikely configurations. *If we wait long enough*, the right kind of configuration will arise—the kind of conditions necessary to give rise to an entirely new universe. A brief fluctuation of matter and energy may accumulate in a small region, distorting the fabric of space-time in just the right way to pinch off and create a disconnected bubble of space. That bubble can expand and grow, ultimately cooling off and condensing into stars and galaxies. It could be the universe in which we live.

We may, in other words, be able to explain the apparently fine-tuned features of times near the bang by appealing to conditions before the bang. It may be that all of the particles we see were originally collected into a small dense region because it is easier to create a new bubble universe in such a configuration than it is to make a large, dilute universe from scratch. The growth of entropy in our observable universe, and the corresponding arrow of time, may be a reflection of the larger multiverse's insatiable desire to create ever more entropy by giving birth to new baby universes. If we could just have an angel's-eye view of the entire ensemble, it might all look quite natural.

Or . . . perhaps not. Once again, we lack the wherewithal to put such ideas to the test. But sometimes we have to go out on a limb, proposing and refining ideas before we understand them well enough to know how to test them. We don't know until we put in the effort to understand where these ideas lead us. The universe is certainly trying to tell us something; it's our job to attempt to make out what it's saying as best we can.

STEPHON H. S. ALEXANDER

is an associate professor of physics at Haverford College. He received his Ph.D. in physics from Brown University in 2000, with a dissertation titled "Topics at the Interface between String Theory and Cosmology." From 2000 to 2002, he held a postdoctoral fellowship from PPARC (the Particle Physics and Astronomy Research Council of the United Kingdom). He recently won the National Science Foundation CAREER Award and was elected a National Geographic Emerging Explorer. His research focuses on unresolved problems—such as the cosmological-constant, or dark-energy, problem—that connect cosmology to quantum gravity and the standard model of elementary particles. In particular, he uses observations in cosmology to both construct and test fundamental theories.

JUST WHAT IS DARK ENERGY?

STEPHON H. S. ALEXANDER

Physicists know that all natural phenomena—and most of our advanced technology, such as cell phones and global positioning satellites—are rooted in two physical principles, quantum mechanics and the principle of relativity. These two, originally not seen as connected, have been shown to be so, in powerful and surprising ways. In the late 1920s, Paul Dirac and others developed the concept of the quantum field, which shows how the combining of Einstein's special relativity with quantum mechanics accounts for the unification of three of the four fundamental forces of nature—the electromagnetic force and the weak and strong nuclear forces.

Despite this success, a unification of quantum mechanics and Einstein's general theory of relativity—which describes gravity, the fourth fundamental force—has yet to be discovered. Once students of physics become convinced of the far-ranging impact and magic of general relativity and quantum mechanics, they can't help but imagine a deep connection between the two. Some of these ambitious students may even decide to spend their careers attempting to unify all four forces. However, in this essay, I will argue that—beyond aesthetics— the physics community has been forced to search for unification because of the existence of a strange form of invisible energy known as *dark energy*, which influences both realms.

Today physics is pretty much where it was at the turn of the last century, just before the dawn of quantum theory and relativity. Back then, many prominent physicists regarded their experiments as the sweeping up of minor details; they were convinced that the physical laws that govern the universe had, by and large, been figured out. This attitude was crisply put in 1894 by the first American physics Nobel laureate, Albert Michelson, who said, "It seems probable that most of the grand underlying principles [of physical science] have been firmly established." As it happened, quantum mechanics and general relativity had to be invented to explain those "minor" experimental details. The dark-energy problem challenges today's physicists to learn from history and avoid making similar presumptions.

The most precise physical theory to date is quantum electrodynamics (QED), Richard Feynman's formulation of quantum field theory. QED is a perfect marriage between quantum mechanics and Einstein's special theory of relativity. The equations of QED can spit out a tiny value of the magnetic field generated by the quantum spin of the electron that agrees with measurement to within nine decimal points (equivalent to accurately measuring the distance between New York and Los Angeles to within the thickness of a human hair). But its mother theory, quantum field theory (QFT), incorporating as it does all of physics (exluding gravity), from distance scales trillions of times smaller than a centimeter all the way up to the behavior of complex molecules, requires the existence of a form of energy—dark energy—and makes a prediction that, beyond 120 decimal points, *disagrees* with observations!

Dark energy, itself directly unobservable, is the most bewildering substance known—the only "stuff" that acts both on subatomic scales and across the largest distances in the cosmos. We find evidence of its activity deep within atomic nuclei and in the motion of distant stars. Its ubiquity captured the imagination of great minds, including that of Einstein himself. Because it is so elusive, dark energy goes by a few aliases, including the *cosmological constant* and *vacuum energy* (I will use these names interchangeably). This essay will display the surprising connection between the smallest and largest scales in our universe.

I want you to picture the empty space that pervades the universe as an ocean. We know that tiny waves on the sea can build into tsunamis. Dark energy is a substance that has this wavelike property in the very fabric of space. At subatomic distances, these dark-energy waves are created by quantum phenomena I will describe after discussing how dark energy shows its face in the heavenward, or cosmological, manifestation of Einstein's general theory of relativity.

Relativistic Cosmology

The general theory of relativity tells us that space and time do not together constitute a fixed, empty stage where matter and energy can dance about; rather, space-time itself bends in reaction to matter.

Why should space behave this way?

Einstein showed us why by questioning the two different ways in which mass can arise—that is, either by acceleration or by gravitational attraction. Let's revisit Einstein's *Gedanken* (thought) experiment involving two people in separate elevators who cannot observe their outside surroundings. One elevator passenger, unbeknownst to him, is in outer space. At first he is floating freely, but once the elevator starts to accelerate upward, his feet come down to the elevator floor, giving him a sensation of weight (mass). The other rider, confined to an elevator at rest on Earth, has the same experience of weight, in this instance because of Earth's gravitational attraction. This imagined experiment led Einstein to the equivalence principle, which he called "the happiest thought of my life." The equivalence principle says that accelerated motion in a nongravitational environment is the same as being at rest in the presence of a gravitational field. Two completely different states of motion give the same impression of mass. Let's explore the ramifications.

Einstein realized that there was a new reality behind the relativity of motion, illuminating the mystery of the nature of gravity: mass and gravitational force are both manifestations of the bending of space. Just as his special relativity showed the constancy of the speed of light in all frames of reference, the general theory of relativity states that whereas all states of motion (including rest) are relative, what is absolute is that space-time itself can be bent in the presence of matter or energy.

The experience of both elevator riders can be explained by the equivalence principle. The energy involved in moving the elevator curves space and thus causes the elevator to accelerate. Likewise, the mass of the Earth bends space, creating an attractive gravitational force. The ultimate consequence of general relativity is that gravity operates without the need for a background space; rather, it *defines* the background space. In contrast, consider the magnetic force, which is exerted by a magnetic field emanating from two magnets and requires space in which to extend. The gravitational field, likewise, carries the gravitational force, but unlike the other fields in nature, it does not require a background space-time; it is itself the background space-time.

The general theory of relativity applies to the behavior of space and time throughout the universe, just as the speed of light is the same for observers everywhere. Using general relativity, physicists have sought to understand the structure and behavior of space-time in the presence of matter and energy. The space-time structure that has emerged influences the movement of matter and light; in some cases, such as a black hole, space can warp so much that light is unable to escape. General relativity has enjoyed unprecedented success in both explaining and predicting new features of gravity, all of which have been observed. But there remains one dark problem that is still being grappled with.

Gedanken Cosmology

Here is a thought experiment. Let's think of the space of our universe as the surface of a gigantic balloon. Every point on its surface is the same as every other point. Imagine that every galaxy corresponds to a static point on the surface of the balloon universe. If we know the distribution of matter on the points of the balloon, the equations of general relativity will tell us how space on its surface changes. By the simple assumption of the Copernican principle—that matter in the universe is evenly distributed and there are no preferred observers— Einstein and, independently, the Russian mathematician Alexander Friedmann were able to find a solution using general relativity that explains the large-scale behavior of the universe's space-time. They found that the matter in the universe has the effect of someone blowing air into the balloon: it expands. Their solution of the equations was compatible with the Copernican principle, but, astonishingly, the static nature (and observational bias) of space was refuted. At that time, the expansion of the universe had not been observed, so their solution seemed to make the wrong prediction.

Every good physicist knows how to save the general theory of relativity in such a case—which is to insert a "fudge factor" that preserves the good parts of the theory while eliminating unwanted aspects. Einstein realized he could introduce a constant—the cosmological constant—into his equations to counterbalance expansion or contraction and render the universe static. In 1924 the astronomer Edwin Hubble showed that the universe was indeed expanding, and Einstein happily got rid of the cosmological constant, which he called his "greatest blunder." Little did he know that this fudge factor, when realized physically, behaves exactly like an invisible and strangely repellent fluid evenly filling the universe: it is our elusive dark energy.

The expanding universe described by the Einstein equations correctly predicts that the universe is some fourteen billion years old. At zero seconds, the universe emerged from a singularity into a dense, radiation-dominated epoch, and as the universe cooled, the radiation transmuted into the light elements (hydrogen through lithium), which coalesced into the first stars and galaxies. This relativistic model of cosmology became known as the standard big bang scenario, but despite its great successes, it could not explain how the smooth Copernican principle underlies the variegated structure—galaxies and galaxy groups—we see in the night sky. But at least the cosmological constant went away, for the time being.

Then, in 1998, cosmologists Saul Perlmutter, Adam Riess, and Brian Schmidt looked at the motion of the distant exploding stars known as supernovas and realized that Hubble's observations of a constantly expanding universe were incorrect at enormous distances—distances greater than Hubble had been able to observe. They found that the universe's expansion was accelerating. Attempting to explain what sort of agent would cause this acceleration, cosmologists quickly found the culprit: it was the cosmological constant. In fact, the cosmological constant seems to be the only agent that can explain both the observations of Perlmutter et al. and a number of other observational nuisances, such as galaxy formation and persistence. The cosmological constant, now better known as dark energy, was no longer a fudge factor, and its true origin and nature had to be confronted by all of physics.

At large distance scales, the electromagnetic force and the two nuclear forces are irrelevant compared with the gravitational force. We now have observational evidence that dark energy—an omnipresent, evenly distributed, repellent substance—acts on gravity to make the very fabric of space accelerate. The problem is that there is no other substance known to us in nature that has this property of making space itself repellent. Dark energy shares a universal quantum feature of all subatomic particles—a feature that endows it with negative pressure and positive energy. (Negative pressure can be thought of as no different from the rebounding of a trampoline that propels a child into the air.)

However, the general theory of relativity does not satisfactorily reveal the nature and origin of dark energy. All it says, so far, is that empty space can have a constant energy and that this constant energy would constitute the repellent force accelerating the universe's expansion. General relativity, though, is happy with or without the cosmological constant. Our most precise understanding of matter and energy is in the subatomic realm of quantum field theory. The notion of emptiness, or the vacuum, in QFT can help identify the source of that constant energy.

Feeling the Field

We are more familiar with fields than with anything else in nature; we hear, see, and feel them. Light, sound, and heat are all best described as fields because of their consistency over a particular region. Because of their extended nature, fields are the agents that carry all four of the fundamental forces of nature. For example, an electric field radiates outward from an electrically charged particle to attract an oppositely charged particle. Einstein won the Nobel Prize for showing that light radiation can kick electrons out of a metal; he thus established the particle-like nature of the electromagnetic field. So, fields can be particles. But can particles be fields?

Another thought experiment. Imagine a river whose flow is interrupted by a waterfall. Most of the water going over the fall is part of a continuous stream, but often drops of water will break apart from the stream. A small fraction will evaporate, but most will rejoin the stream at the bottom of the fall. QFT is analogous to the waterfall. Let's think of the electron itself as a field, living everywhere in space. Like the water droplet, the electron will eventually return, as a particle, to its parent electromagnetic field.

When quantum mechanics is unified with special relativity, quantum field theory makes two important predictions. First, fields are the fundamental objects that live in empty space (that is, space in which there are no particles). We can think of the river as the vacuum state the fields are living in. Particles can be created from or (appear to be) destroyed in the vacuum, just as the droplets of water left and returned to the waterfall. Paul Dirac discovered a stunning consequence of QFT: antimatter. He reasoned that since the vacuum state is the lowest-energy state possible, a particle cannot spontaneously be created unless its antiparticle is also created.

It has been shown that the spontaneous bubbling of matter and antimatter in the vacuum generates an electric force that causes two metal plates to attract each other. This quantum feature of the vacuum is called the Casimir effect, and it is direct proof of the existence of vacuum energy. Now for the punch line: the properties of the vacuum energy are none other than the quantum manifestation of dark energy. When we consider this energy as existing at every point in empty space, we are forced to conclude that it contributes a near infinite amount of cosmological constant to the universe, which would cause the universe to instantaneously inflate away, at a rate far beyond that which Perlmutter and his colleagues observed. The vacuum energy is unbounded, according to quantum field theory, but we can subtract it away without negating the useful predictions of nuclear physics, where QFT works almost perfectly.

The ability to subtract away the nearly infinite vacuum energy in QFT is analogous to lowering the sea level. If you're in a boat in the middle of the ocean, you are aware of the rise and fall of the ocean waves. If the world's oceans were all steadily draining away, you would not be aware of that. You would still be aware only of the surface-wave action on your boat. The huge amount of vacuum energy, like the relative height of the seas, can be subtracted away without affecting the good predictions and experimental successes of QFT. Any effects of vacuum energy we measure in the lab, like the waves rocking our boat, require this infinite subtraction.

Recall that general relativity says that all energy and matter, regardless of their form, impact the fabric of space-time. Thus it is with dark energy, which must also act on gravity to create more and more empty space at an ever greater rate. Our currently accelerating universe is consistent with this expectation. While QFT, our most successful and technologically useful description of subatomic physics, shows us why dark energy must exist, it makes that embarrassing prediction for gravity. Or could it be that QFT is really correct? Could it be that the vacuum energy is indeed nearly infinite and gravity is just insensitive to it? If this is so, then there is something missing in general relativity. Another possibility is that quantum field theory and general relativity are both incorrect in describing dark energy and the structure of the vacuum.

Physicists have wrestled with the dark-energy/cosmological-constant problem for the last half century. Progress has been slow, although there are promising proposals for a resolution. I believe that we need to tackle the two underlying principles of general relativity and quantization impartially to find a resolution.

New Directions in the Void

Most physicists who have been humbled by this problem agree that a resolution will most likely require a dramatic shift in our current paradigm. There are some logical directions that have been taken that are worth mentioning.

1. QFT is fine. Change gravity.

Given that we have measured dark energy on subatomic scales, we could conclude that general relativity is inadequate, since it does not have quantum physics built into its structure. In quantum terms, the vacuum is the absence of all matter and antimatter. In general relativity, there is no distinction between matter and antimatter, and in fact the existence of antimatter doesn't even make sense; only the relationship between matter, energy, and curvature makes sense—at least macroscopically. But Canadian physicist William George Unruh proved that the vacuum state in QFT is dependent on an observer's state of motion; one observer's vacuum state of no matter is another's soup of matter. The QFT vacuum is thus in conflict with general relativity, which puts all observers' states of motion on equal physical footing.

It could be that a quantum version of gravity would naturally be "blind" to the tremendous amount of dark energy generated by the constant activity of quantum fields. Theorists have tried to cook up quantum modifications to general relativity that render it blind to vacuum energy—modifications that generate a "screening" effect—but they all lead to pathologies at cosmological distances, and we observe the effects of dark energy at those distances as well.

2. Dark energy—act as if your life depended on it.

By replacing the fundamental pointlike structure of matter with vibrating strings of energy, string theory, remarkably, is able to unify QFT with gravity. In string theory, mathematical solutions of the strings give different vibrational patterns associated with the different forms of matter and energy— including the graviton, the particle associated with gravity. However, this compelling picture of ultimate unification comes at a price: a potential paradigm shift with regard to how we should think about dark energy, our own place in the universe, and even what we mean by the word *universe*.

Aside from being a (mostly) successful unification of general relativity and quantum mechanics, string theory admits a countable infinity of vacuum solutions, each giving different values of dark energy for different universes in a multiverse. Some leading string theorists argue that if the multiverse is so big that all of these solutions are realized in vastly distant parts of space, then we could be living in one of those pockets. The Nobel laureate Steven Weinberg has shown that if dark energy were too large, its repulsive nature would have prevented the light elements present immediately after the big bang from collapsing into stars and galaxies. And yet collapse they did. This argument dovetails with the anthropic principle: we just happen to live in a part of the multiverse where the amount of dark energy is just right for life to exist. This principle may well find a home in string theory and will modify how we think about the vacuum—different vacuums may yield different universes with different physical laws. String theorists and mathematicians are hard at work on concretizing this proposal.

A Future Direction: Relative Reductionism

A great lesson in the evolution of physical theories is that new paradigms cover a broader range than their predecessor theories. After all, Newton's laws are a perfectly fine description of gravity for most observable situations. They emerge from general relativity, whose effects cannot be seen in our solar system—except for a tiny "anomaly" in the motion of Mercury due to its close proximity to the sun, whose gravitational field is very large. The same holds in quantum field theory. Imagine the quanta of fields as pixels on a universal computer screen; just as we ignore the pixels when looking at the screen, QFT gives us the correct picture of macroscopic bodies, like trees and cows and automobiles, at macroscopic distance scales.

One pattern that becomes clear is that the more we view

matter and energy as subject to relativity, the more new fundamental particles emerge. But are these particles fundamental, indivisible entities? In maintaining our bias toward reductionism—i. e., the conviction that these new particles are indivisible—we maintain the intrinsic divide between matter and space-time. But dark energy is telling us that we have to confront this bias—that we have to see the vacuum as relative and not as an absolute entity; we need to relativize the quantum notion of emptiness.

Nature has given us a hint. Dark energy operates over large distances, where the effects of gravity dominate, but paradoxically, according to QFT, it should be nearly infinite at subatomic scales, where gravity seems to have no effect at all. As I've noted, at smaller and smaller distance scales we discover more and more new fundamental particles. But at macroscopic scales, quantum effects can do the reverse—that is, fundamental particles (electrons, say) can collectively behave like a new entity, which we call an "emergent quantum phenomenon." Superconductivity and superfluidity are examples; in these emergent physical phenomena, the collective behavior is what is fundamental.

In line with what Einstein taught us, the universal vacuum would have to be consistent for any observer's state of motion, while also displaying the mass-energy and space-time curvature of general relativity that we observe. However, observers are themselves matter, created from this empty space. The definition of what makes up an observer differs in QFT and in general relativity. There has always been a silent tension between these two paradigms in physical law: emergence and reductionism. If observers, like our elevator riders, were just different aspects of a new physics, could the dark-energy problem go away?

In this view, the notion of a fundamental particle—or of an observer—would not be absolute. Perhaps what are thought to be the fundamental particles we see in nature on subatomic scales are the equivalent of a quantum emergence of the fabric of space itself. What is a fundamental particle to one observer is emergent to another. In this view, a new relativity principle would make not just motion but also the very notion of a fundamental particle (and its associated vacuum) relative. A physical consequence would be that matter can create space and space may curve itself into matter. The nearly infinite amount of dark energy we expect at subatomic scales might actually be an artifact of emergent space-time matter over larger distances. Such a possibility may even end up explaining another observed mystery, which has puzzled astronomers about galaxies for the past half century: the invisible ("dark") matter that must be present to keep them from flying apart.

SARAH-JAYNE BLAKEMORE
received a bachelor's degree with honors in experimental psychology from Oxford University and her doctorate in 2000 from University College London. She is currently a Royal Society University Research Fellow at the Institute of Cognitive Neuroscience, University College London.

Blakemore's research centers on social cognitive neuroscience. Her group at UCL studies the development of mentalizing, action understanding, and executive function during adolescence, using a variety of behavioral and neuroimaging methods. A second focus of their research is on social cognitive deficits in autism. Blakemore is coauthor (with Uta Frith) of The Learning Brain: Lessons for Education.

DEVELOPMENT OF THE SOCIAL BRAIN IN ADOLESCENCE

SARAH-JAYNE BLAKEMORE

A few decades ago, many people would have found it hard to believe that profound changes in the brain take place after early childhood. Some would even have argued that our brains are largely fixed by the age of three. Now, due to recent research using modern brain-imaging techniques, scientists are discovering that the human brain does indeed change well beyond early childhood. Some brain regions—in particular, the prefrontal cortex (PFC)—continue to develop during adolescence and even beyond. The PFC is involved in a wide range of cognitive abilities, including planning and decision making. It is also part of the *social brain*—that is, the network of brain regions involved in understanding other people.

Much has been known about early brain development since experiments on animals—rodents, cats, and monkeys—carried out in the 1950s and '60s. One major developmental process affects the "wiring" of neurons, or brain cells—that is, the intricate network of synapses constituting the connections between them. Early in the brain's development, it begins to undergo synaptogenesis, the formation of new synapses; remarkably, the number of such connections in a baby's brain greatly exceeds adult levels. This is followed by a period of synaptic elimination—or pruning—in which excess connections wither away.

The animal research showed that synaptic pruning can be influenced by the particular environment an animal experiences. Synapses that are frequently used are strengthened, while infrequently used synapses are eliminated. The pruning of synapses is much like the pruning of a rose bush: eliminating the weak branches allows the remaining branches to grow stronger. Research around the same time also indicated that there are critical periods of brain development early in life during which an animal needs to be exposed to sensory stimulation for normal brain development to occur. The early animal experiments showed that these processes are mostly over by about age three, at least in sensory regions of the monkey brain.

On the basis of this research, textbooks often suggest that the crucial phase of brain development in humans occurs in the first three years of life and that during this time children should therefore be exposed to all sorts of learning experiences. However, this argument neglects the fact that monkeys are sexually mature by age three and do not go through the same extended developmental period as humans do. Indeed, research carried out in the 1970s and '80s demonstrated that the time period for both synaptogenesis and synaptic pruning is not the same in humans as it is in monkeys.

In the 1970s, Peter Huttenlocher, at the University of

Chicago, collected postmortem brains from humans of all ages and found that the frontal cortex in children was remarkably different from that in adolescents. Whereas in sensory brain areas the number of synapses reaches mature levels by mid-childhood, the number of synapses in the prefrontal cortex continues to increase and then declines during adolescence.

Huttenlocher's findings have been confirmed by recent research on living human subjects, using such noninvasive brain-imaging techniques as magnetic resonance imaging (MRI). These studies have indicated that several regions of the cortex, and especially the PFC, undergo substantial changes throughout the first two or three decades of life.[1]

Two main developmental changes have been detected in these MRI scans. First, there is an increase in the volume of white matter in some brain regions, including the PFC, and this increase continues throughout adolescence and into the twenties. White matter consists of axons, the long fibers that carry the electrical signals from one neuron to another. Axons appear white because they are coated in a white fatty substance called myelin, which acts as an insulator and speeds up the transmission of signals down the axon. The implication is that axons continue to accumulate myelin for several decades, thereby speeding up processing in the relevant brain regions.

A second change in the human brain during adolescence is a decrease in the volume of gray matter in the prefrontal cortex. Gray matter consists of the cell bodies of neurons and their tentacle-like connections forming synapses with other neurons,

[1] A. W. Toga et al., "Mapping Brain Maturation," *Trends in Neuroscience* 29, no. 3(2006): 148–59.

and the pattern of gray-matter development has been interpreted as reflecting changes in the number of synapses in a particular brain region.

Implications for Teenagers

What are the implications of this continuing brain development for adolescents? Adolescence is a period of life characterized by change—the period of physical, psychological,
and social transition between childhood and adulthood. At
the beginning of adolescence, around the onset of puberty,
there are dramatic changes in hormone levels and consequent alterations in physical appearance. This period of life
is also characterized by psychological changes related to
mood, self-consciousness, one's sense of identity, and one's relationships with others. But the recent neuroscience studies
just described suggest that hormones alone do not account
for these psychological changes. What are the implications
of the development of the prefrontal cortex for social
cognition?

Mentalizing

Understanding other people involves reading their behavior in
terms of their underlying mental states, such as intentions and
desires, a process called mentalizing. Brain-imaging experiments and studies of patients with brain lesions have shown

that mentalizing relies on the network of regions known as the social brain.[2]

There is a rich literature on the early development of mentalizing. Signs of social competence—such as face recognition and recognition of others' emotions—develop during infancy. These early social abilities precede full-fledged mentalizing, such as the ability to recognize others' beliefs, including their false beliefs. In a typical false-belief study, the child is asked where "Sally" will look for a toy that she left in a certain place and that was subsequently moved by "Anne" when Sally was out of the room. Understanding that Sally will not know the toy's new location depends on the test subject's ability to distinguish between Sally's false belief and reality. This is achieved in most children by age four or five.[3]

While the early development of mentalizing has been well studied, there has been surprisingly little empirical research on social cognitive development beyond childhood, perhaps because most children can pass even quite complex mentalizing tasks by four or five. Yet the brain structures that underlie mentalizing (including the medial PFC) undergo substantial development well beyond early childhood.

A small number of recent brain-imaging studies have looked at the development of the functioning of the social brain during adolescence. There is some indication that activity in the medial prefrontal cortex during mentalizing tasks decreases between the onset of adolescence and the beginning

[2] C. D. Frith and U. Frith, "Social Cognition in Humans," *Current Biology* 17, no. 16(2007): 724–32.
[3] J. Barresi and C. Moore, "Intentional Relations and Social Understanding," *Behavioural and Brain Sciences* 19(1996): 107–54.

of adulthood. For example, a recent functional MRI (fMRI) study investigated the development of communicative intent using an irony-comprehension task. The premise was that, just as understanding others' mental states requires the ability to decouple belief from reality, understanding irony requires separating the literal from the intended meaning of a comment. A dozen adults ranging in age from twenty-three to thirty-three and a dozen children ranging in age from nine to fourteen were scanned. The medial PFC in the children was more engaged than the medial PFC in the adults during this task. The authors interpreted the increased medial prefrontal activity in children as reflecting the need to integrate several cues to resolve the discrepancy between the literal and intended meaning of an ironic remark.[4]

A similar region of the medial prefrontal cortex was more highly activated by children than by adults in a recent fMRI study that involved thinking about one's own intentions. Thinking about your own (or someone else's) intentions to act requires mentalizing. A group of nineteen female adolescents (age range: twelve to eighteen) and a group of female adults (twenty-two to thirty-eight) were presented with scenarios about intentions and actions (e.g., "You want to find out what's on at the theater. Do you look in a newspaper?"). The medial PFC was more active in the adolescents than in the adults when thinking about intentions. Activity in this region decreased with age, which suggests that the neural strategy for

[4] A. T. Wang et al., "Developmental Changes in the Neural Basis of Interpreting Communicative Intent," *Social Cognitive and Affective Neuroscience* 1(2006): 107–21.

thinking about intentions changes between adolescence and adulthood.[5]

The decrease in activity in the medial PFC during adolescence may happen because adolescence is a time when the prefrontal cortex is being fine-tuned by synaptic pruning, and a decreasing amount of activity is therefore necessary to carry out the task in question. An alternative (or additional) explanation is that there is a change in the cognitive strategy for mentalizing that results in the recruitment of other brain regions. Whether decreasing activity during adolescence is due to a new mentalizing strategy or to pruned synaptic connections remains to be seen.

Implications for Society

What are the implications of this research for society?

First, it casts doubt on the notion that extra stimulation for babies and early formal education is necessary for optimal brain development; this research shows that the human brain continues to be malleable for decades.

Second, the demonstration of marked transformations in the brain during adolescence suggests that hormones alone do not account for typical teenage behaviors.

Third, it is possible that the synaptic reorganization taking place during adolescence is influenced by environment, just as

[5] S. J. Blakemore et al., "Adolescent Development of the Neural Circuitry for Thinking About Intentions," *Social Cognitive and Affective Neuroscience* 2, no. 2(2007): 130–39.

early synaptic pruning is. At the moment this is pure speculation, but if it is the case, there are implications for what kinds of experiences adolescents should encounter. Recent research reported in the *British Journal of Psychiatry* by Robin Murray's group at the Institute of Psychiatry, King's College London, suggests that teenagers who smoke marijuana regularly are more likely to develop schizophrenia in early adulthood than those who do not. One possibility is that marijuana affects brain development during the teenage years.

Furthermore, there are likely to be gender differences in social cognitive development during adolescence. How brain development interacts with hormonal change, and how this interaction affects social cognition, have not yet been evaluated. The development of the adolescent brain, a new and rapidly expanding field in neuroscience, is only beginning to be understood.

JASON P. MITCHELL

is principal investigator of Harvard University's Social Cognitive and Affective Neuroscience Laboratory, where he uses functional neuroimaging (fMRI) and behavioral methods to study how perceivers infer the thoughts, feelings, and opinions of others. Mitchell received his B.A. and M.S. degrees from Yale University in 1997 and his Ph.D. from Harvard University in 2003. He is currently an assistant professor in Harvard's Department of Psychology. He has also been a visiting professor at Dartmouth College and Columbia University. He is consulting editor for the journal Social Cognitive and Affective Neuroscience *and associate editor at* NeuroImage *and* Cortex.

WATCHING MINDS INTERACT

JASON P. MITCHELL

To anyone who hasn't yet met one, the members of *Homo sapiens* must surely seem like the species least likely to dominate the planet. Compared with most other terrestrial mammals, humans are a pretty scrawny bunch. We don't come equipped with sharp claws or fangs; we're not particularly swift or strong; we don't have special physical adaptations that would allow us to fly, poison potential predators, or cloak ourselves in camouflage. To a hungry lion, this hairless, frail ape, standing upright on the savannah, must have been the least troublesome of afternoon snacks.

And yet, despite these obvious physical shortcomings, humans are Earth's undisputed masters—at least for the time being. We have co-opted the life history of hundreds of plant and animal species through domestication, and we are responsible for the extinction of hundreds of others. Our technology continues to transform the planet itself—the land, the sea, the atmosphere. Our command of Earth is so nearly complete that humans maintain a presence in places that harbor few or no other organisms, such as the South Pole and low-Earth orbit.

How has our species of fragile-seeming primates managed to subjugate the rest of Earth's inhabitants? The answer, of course, is that natural selection has equipped us with an adaptation more fearsome than teeth or claws: the human brain.

Composed of 100 billion neurons, each communicating with an average of 1,000 other neurons dozens of times per second, the adult human brain is the most complex object in the known universe, a biological supercomputer with a processing capacity that dwarfs the most sophisticated silicon-chip computers we are likely to see in our lifetimes. Thanks to the tremendous computational power of our brains, humans have been able to transcend the need for the kinds of weapons in which many other organisms invest.

And yet, how exactly do our prodigious brains provide an advantage to our species? Unlike claws and wings and venomed fangs, our brains have no direct interaction with the environment around us. (It's safe to say that things have gone horribly wrong if you suddenly find your brain in contact with the objects around you.) Instead, our brains are the prima donnas of the biological world, sequestered behind a bony fortification half an inch thick, greedily devouring more than 20 percent of the body's energy. Since brains do not directly effect change in the world around them, how did they give us an adaptive leg up on other organisms?

Unlike most other adaptations, the brain requires a system to translate the only physical actions of which it is capable—the electrochemical firing of its neurons—into the physical actions, such as speech and toolmaking, that have enabled us to dominate our surroundings. In the twenty-first century, most of us are familiar with an analogous translational system: the operating systems and other software that translate the physical actions of a computer processor (the electrical switching of binary diodes) into the physical actions that make computers so useful—for example, the arrangement of colored points on a monitor into meaningful images, such as words or pictures. The particular system of translation used by the human brain is known as the *mind*, which may be defined as the algorithms by which one set of physical actions is mapped onto a different set of physical actions by the brain.

The branch of science that attempts to describe these algorithms is psychology, whose ultimate goal is the assembling of a full catalog of the subroutines and other processing steps with which human beings translate physical information from the environment (photons of light, vibrating waves of air) into a physical representation (neural signaling) inside their heads and then back into physical action on the environment (movements, speech, and other observable behavior).

Take vision, for example. The human perceptual system comprises a set of processing steps that somehow transform a two-dimensional pattern of photons on the retina into a neural experience of colored, textured objects existing around us in three-dimensional space. It has taken us some time to realize exactly how difficult and computationally challenging the investigation of these basic functions is. In an oft-told (possibly apocryphal) tale, Marvin Minsky, one of the founders of the field of artificial intelligence, assigned the problem of computer vision to a student as a summer project. Thirty years later, we are still struggling to identify the translational algorithms that guide basic perceptual and motor systems, as well as those behaviors that appear uniquely part of the human mental repertoire, such as language, reasoning, and forms of social thought.

Until about twenty years ago, most of our knowledge about the way the brain produces the mind relied on studies of patients with specific brain damage from a stroke or head injury, or on tentative extrapolation from experiments on non-human animals. However, recently developed techniques for functional imaging and virtual lesioning have finally allowed researchers to examine the living, healthy human brain as it goes about its business of processing information. Methods such as functional magnetic resonance imaging (fMRI), its predecessor, positron emission tomography (PET), and newer techniques such as near-infrared spectroscopy (NIRS) now allow researchers to pinpoint areas in which metabolic activity—for example, greater blood flow—increases when participants perform a particular task. In addition, more exotic techniques—such as transcranial magnetic stimulation (TMS), in which rapidly changing magnetic fields are used to create short-lived electrical disruptions of cortical function (virtual brain lesions)—enable researchers to catalog the behaviors that humans can and cannot perform when the functions served by a particular brain region are temporarily disabled. These techniques have dramatically accelerated our scientific understanding of the human mind, permitting a direct look at the underlying hardware on which our mental algorithms are being run.

Two assumptions about the brain allow these new brain-imaging methods to inform the study of human cognition.

First, we generally presume that different brain regions have different processing portfolios—that is, that different brain regions are likely to serve distinct subroutines of the mind. Second, brain regions are thought to serve a limited number of processing operations, perhaps only a single specific type of computation. These two working assumptions allow powerful insights into the organization of the mind.

Imagine that a researcher is interested in the question of whether perceiving and identifying the faces of other people ("That's John Travolta!") relies on the same types of information processing as perceiving and identifying inanimate objects ("Is that a white leisure suit he's wearing?"). As it turns out, these two mental abilities engage neighboring, but distinct, brain regions, suggesting that different cognitive processing— different transformations of incoming information—takes place between the point at which photons impinge on one's retina and the point at which one has the conscious experience of perceiving either a face or an inanimate object. If, on the other hand, face and object perception relied on the operation of the same brain regions, researchers would have probable cause to infer that the same kind of information processing mediated between the retina and an understanding of what one was looking at. If either of these two assumptions turns out to be untrue—for example, if different brain regions often serve the exact same mental operations—the use of brain imaging to inform psychology will prove severely limited. Fortunately, thus far, psychologists have no reason to doubt that the brain is organized like other machines, composed of parts that perform specialized and distinct functions.

In this way, neuroimaging has unearthed some unexpected

features of human psychology—features once underappreciated. For example, it seems that our recall of past events, rather than being a single, monolithic operation, can be deconstructed into several distinct processing mechanisms. The brain regions that contribute to memory vary depending on what kind of information is to be remembered (faces, words, locations), whether one needs to recall specific details of an event or merely the gist of it, and how long ago the memory was encoded. Conversely, neuroimaging studies suggest that some mental experiences we think of as being distinct from one another actually rely on the same information-processing circuits: for example, imagining an object (your cat's ears, say—are they pointy or floppy?) engages some of the same visual-processing areas that serve real-world perception—that is, actually seeing a cat in front of you. By demonstrating that some mental experiences feel unitary but comprise multiple processes, whereas others feel distinct but rely on the same processing areas, neuroimaging has forced psychologists to reconsider many of the intuitive divisions and theoretical constructs that developed before we could so easily examine the neural hardware on which the mind runs.

Perhaps the least anticipated contribution of brain imaging to psychological science has been a sudden appreciation for the centrality of social thought to the human mental repertoire. Although *Homo sapiens* surely owes much of its evolutionary success to our unparalleled ability to reason in flexible and novel ways, the most dramatic innovation introduced with the rollout of our species is not the prowess of *individual* minds but the ability to harness that power across many individuals. Indeed, the scale and complexity of human behavior—including our current world dominance—is supported in large part by mechanisms that allow us to coordinate large groups of people to achieve goals that individuals could not. Consider the number of people required to design, construct, and operate an airplane, erect a building, or run a national government.

In order to integrate the behavior of many individuals, the human mind must be capable of at least two kinds of special processing. First, to have any hope of coordinating the minds of others, we must have a way to understand what's happening inside them—that is, a set of processes for inferring what those around us are thinking and feeling; what their goals, desires, and preferences might be; and what personality traits and temperament differentiate them from other people. In other words, we have to be "mind readers," capable of perceiving the mental states of the people around us. Second, we must possess tools not only for passively inferring the contents of others' minds but also for actively influencing what others think and feel. The surest way to enable one to know what's on another person's mind is to implant one's own thoughts and feelings into the other's mind, and humans have unique and remarkably powerful means of doing this. Human language can be considered primarily a vehicle for transferring one's own mental states into another mind. Humans are the only animals that explicitly attempt to affect the content of others' minds through direct instruction. Although other primates may try to manipulate the mental states of their conspecifics (for example, through deception), they do not appear to attempt to reproduce their own mental states in other minds.

The emerging field of social neuroscience has suggested that these interpersonal abilities draw on several unexpected characteristics of the human brain. Natural selection has designed specialized regions whose activity appears to be specifically dedicated to the task of understanding the goings-on of other people's minds. Dozens of neuroimaging studies have examined the pattern of neural activity that differentiates between tasks that require the consideration of another person's mental state (How happy does this person look in the photograph?) and those that require consideration of the non-mental qualities of a stimulus (How symmetrical is this person's face?). These studies have uncovered a set of brain regions that are preferentially engaged during "mind reading": the dorsomedial prefrontal cortex (a region immediately behind your forehead and in line with your nose), the temporo-parietal junction (so named because it is found where the parietal and temporal cortices meet, about two inches above and two inches behind your ears), and the medial parietal cortex (immediately below the crown of your head). In fact, the observation of activity in these regions during social tasks may be the most consistent finding in all of cognitive neuroscience.

Even more intriguing, these brain regions are marked by an unusual property. When people are left to lie quietly in an MRI or PET scanner without any particular task to perform, most of their brain decreases in activity. However, the brain regions identified during mind-reading tasks continue to churn away. This chronic engagement of "social brain" regions suggests that the human brain has a predilection for contemplating the minds of others. Our tendency to anthropomorphize—to see mind where none truly exists, as in inanimate objects or the forces of nature—may well result from the chronic overactivity of those brain regions implicated in social thought. Human

minds seem always at the ready to tackle another mind, a pro-
clivity that may lead us to perceive the world as being chock-full
of other mental agents.

The special neural status of social thought is further suggested by another unusual feature of these brain regions: they
tend to deactivate when a person thinks about something other
than a mind. Brain regions that serve other mental functions do
not generally exhibit such decreased activity when their particular information-processing services are not required. For
example, brain regions involved in arithmetical calculations do
not "shut off" when you think about something other than
numbers; they simply cease responding over and above their
resting rate of metabolism. Not so for brain regions involved in
social thought, which routinely show decreased activity when
not in use. This peculiar—and poorly understood—aspect of
social-brain regions suggests that social thought may be an
activity incompatible with other kinds of information processing. Perhaps our mental algorithms cannot simultaneously
remain vigilant for the presence of other minds and also interact with entities that are inherently mindless (such as mathematical operations or inanimate objects), but must instead
suspend the tendency to approach the world in a social manner
when faced with nonminds. If we were unable to suppress the
predisposition to see all things as having a mind, think how
hard-hearted we'd have to be just to pour scalding water into a
mug, pound a nail, or slam-dunk a basketball.

Finally, social neuroscience has begun to demonstrate just
how exquisitely sensitive our minds are to the goings-on of the
minds around us by suggesting that our brains spontaneously
mirror the pattern of activity of other brains in our vicinity.

When we see a person who looks afraid, our brain responds in the same way it does when we experience fear ourselves—by activating a small region in the brain known as the amygdala. When we see someone about to slam a door on his finger or being injected with a syringe, our brain responds as though we ourselves were experiencing pain—by activating the anterior cingulate cortex. People also often wince in virtual pain when they see or contemplate others undergoing such experiences. If we were to see another person take a deep whiff of rotting garbage, our brains would respond as though we ourselves were disgusted, by activating the insula. And if we watch another person try to achieve a particular goal (such as picking up a desired object), our brains respond as though we were doing the same, by activating so-called mirror areas in the parietal and frontal lobes. Such observations suggest that the human mind naturally attempts to engage in the same kinds of information processing as neighboring minds—that our brains prefer to be in register with the brains around us. Although the implications of this nascent discovery have yet to be explored in full, our drive to get on the same mental page as other people hints at a hidden complexity in human social interactions that includes our brains simultaneously struggling to figure out and to adopt the same processing states as those of others.

These observations about the brain's social proclivities have refocused a good deal of psychological research on the problem of just how the mind engages with other minds. Successfully interacting with, predicting, or influencing the mind of another person requires an extraordinary set of cognitive skills. What kinds of information-translation processes transform a partner's raised eyebrow or sideways glance into an understanding of that person's thoughts and feelings? What processes transform one's mental states into complex utterances, complete with a pragmatic awareness of what a listener can comprehend (as shown by, for example, differences in the way we speak to children and adults)? We've only just begun to figure out the answers to these questions, but armed with new technologies for imaging the living human brain and a new appreciation for the importance of social thought, psychological scientists may soon be able to unravel the mind's intricate dance as it responds to, influences, and is influenced by other minds.

MATTHEW D. LIEBERMAN,
*an associate professor of psychology at UCLA, received his
Ph.D. in psychology from Harvard University in 1999. His
research interests include such social cognitive neuroscience
topics as self-control, self-awareness, automaticity, social rejection, and persuasion. He has published in numerous journals,
including* Science, Nature Neuroscience, Annual Review
of Psychology, American Psychologist, Journal of Cognitive Neuroscience, *and* Journal of Personality and
Social Psychology, *and his research has been funded by
grants from the National Institute of Mental Health, the
National Science Foundation, the Guggenheim Foundation,
and the Defense Advanced Research Projects Agency. His
work has received coverage in* Time, Scientific American,
Discover, *and multiple BBC documentaries. Lieberman is
the founding editor of the journal* Social Cognitive and
Affective Neuroscience *and was the 2007 recipient of the
APA Distinguished Scientific Award for Early Career Contribution to Psychology. He is currently working on a book
tentatively titled* Experience Shrugged: The Rise of
Simulated Experience in Mental Life and the Modern
World.

WHAT MAKES BIG IDEAS STICKY?

MATTHEW D. LIEBERMAN

In 1641 René Descartes published his *Meditations on First Philosophy*, in which he presented his theory of mind-body dualism, later known simply as Cartesian dualism. According to Descartes, the mind is animated by an immaterial soul distinct from the realm of the physical and all physical processes. There is the mental and there is the physical, and never the twain shall meet (except perhaps through the pineal gland, or perhaps by God's intervention; otherwise it is difficult to explain the nearly perfect correlation between the mind's desire to open a door and the body's simultaneous performance of the desired act). A few decades later, J. J. Becher published *Physica subterranea* (1667), which similarly focused on an invisible entity. Becher proposed that all flammable materials are flammable because they contain phlogiston, a hypothetical substance without color, odor, taste, or weight; thus fire, too, is animated by a seemingly immaterial substance. Descartes' and Becher's ideas were widely discussed and believed in their day.

Times have changed, and so have the fortunes of these two theories. Whereas mind-body dualism is one of the most entrenched ideas of the last millennium, informing policy discussions regarding the ethics of cloning, abortion, euthanasia, and the use of animals in laboratory tests, phlogiston is only occasionally mentioned in scientific circles, and then as a cautionary tale of unscientific theorizing. One might naturally assume that the reason Cartesian dualism endures while phlogiston has fallen out of favor is that the former has garnered scientific support while the latter has been refuted by science. One might assume this, but one would be wrong.

In scientific circles, neither theory is reputable, although scientists still regularly report their findings in dualistic language. One of the fundamental tenets of the modern science of the mind is that the mind is a thoroughly biological and therefore material entity. Moreover, philosophy long ago established that mind-body dualism is logically impossible without the incorporation of numerous convoluted assumptions. Nevertheless, people walk around with an ingrained belief in the simple but implausible form of mind-body dualism that Descartes described. Just consider all the mind-brain and mind-body institutes springing up around the world, all claiming to explore the connection between those entities. Such institutes continue to reify dualism by suggesting that mind and body are distinct enough to need connections. Ongoing discussions of how brain states cause mental states and how meditation uses the mind to alter the brain and body similarly bolster the mind-body distinction.

Why is mind-body dualism a sticky idea that endures in the face of scientific and philosophical disbelief? Why, for that matter, does any idea take hold of large groups of people and endure for decades, or centuries? How do ideas become Big Ideas? Psychologists know a great deal about how the source and content of a message lead an individual to reject or be persuaded by an argument. Malcolm Gladwell's 2000 best seller *The Tipping Point* is a compelling popular account of the kinds of people who serve as an idea-distribution chain, ensuring an idea's wide influence. Most such memes, or contagious cultural ideas, typically come and go in a matter of years, months, or even days. Disco and bell-bottoms may have been cool in the seventies, for all the reasons thought to make ideas persuasive, but come the eighties, disco and bell-bottoms were out and new-wave music and tight jeans were in.

But what about the ideas that are truly enduring, like Cartesian dualism? I argue that Big Ideas sometimes match the structure and function of the human brain such that the brain causes us to see the world in ways that make it virtually impossible not to believe them. I call this explanation the *Deacon doctrine*, in honor of Berkeley anthropologist and neuroscientist Terrence Deacon, who inspired the idea.

In *The Symbolic Species: The Co-Evolution of Language and the Brain* (1998), Deacon provides a counterintuitive account of why humans have come to use language in its modern form. The common account of language use, according to Deacon, is that the human brain evolved *in order to* be able to perform all the mental activities necessary to use the kind of language we use. Deacon turned this logic on its head by suggesting that although the human brain did evolve a capacity for symbolic processing, this was not for the purpose of language per se. Rather, Deacon suggests that it was so that couples could forge a bond of sexual trust that would be respected by the tribe, allowing the men to go off hunting without their mates. It is the next part of Deacon's argument that is critical to the Deacon doctrine. Deacon suggests that language has evolved (and continues to evolve) to fit the structure and function of the human brain, rather than the other way around. He provides extensive evidence that language evolves much more quickly and easily than the brain does, and that as language changes from generation to generation, it almost always changes in ways that make it easier for two-year-old children to learn.

The Deacon doctrine can thus be stated: *one reason Big Ideas are influential and enduring is because they fit with the structure and function of the human brain*. Or, as Deacon puts it, ideas evolve to fit the structure and function of the brain, and as greater fit emerges, the ideas become "stickier." Two effects should be present in cases where the Deacon doctrine applies: first, there should be some form of strong fit between the content of a Big Idea and the structure and function of the brain; second, the Big Idea should have changed over time to better approximate the critical features of brain organization. This essay will consider two Big Ideas for which the Deacon doctrine applies: mind-body dualism and Eastern versus Western culture.

Mind-Body Dualism

Although the scientific consensus is that minds and bodies are made of the same stuff, the science of how the brain makes sense of minds and bodies in daily life is in its infancy. Indeed, nearly all the evidence on the subject is focused on two other topics—making sense of ourselves and making sense of other people—and only incidentally provides a picture of how the brain generates its own mind-body dualism. About a dozen neuroimaging studies, mostly using functional magnetic resonance imaging (fMRI), have found that two regions on the medial (or middle) surface of the brain, one in the prefrontal cortex (medial PFC) and one in the parietal cortex (medial PAC), tend to be more active during introspection—that is, when one is focused on the self, reflecting on one's state, traits, or preferences.[1] Another line of investigation has examined the brain regions involved in recognizing physical indicators of the self, such as visually recognizing one's own face. Somewhat surprisingly, when people are shown pictures of their faces during neuroimaging studies, the medial PFC and medial PAC, the regions involved in focusing on one's nonphysical attributes, are not activated; instead, regions in the lateral PFC and lateral PAC, on the outer surface of the brain, are activated.[2] Additionally, the lateral PAC appears to be involved in observing one's own body movements; disturbances in this region may figure in out-of-body experiences and in the sense—as in schizophrenia—that someone else is controlling one's body.[3]

[1] M. D. Lieberman, "Social Cognitive Neuroscience: A Review of Core Processes," *Annual Review of Psychology* 58(2007): 259–89; W. M. Kelley et al., "Finding the Self? An Event-Related fMRI Study," *Journal of Cognitive Neuroscience* 14(2002): 785–94.

[2] L. Q. Uddin et al., "Self-Face Recognition Activates a Frontoparietal 'Mirror' Network in the Right Hemisphere: An Event-Related fMRI Study," *NeuroImage* 15(2005): 926–35.

[3] O. Blanke et al., "Stimulating Illusory Own-Body Perceptions: The Part of the Brain That Can Induce Out-of-Body Experience Has Been Located,"

A similar distinction can be seen in the brain's processing of other people, depending on whether the test subject is trying to make sense of another person in terms of the mind or the body. When we engage in "mentalizing," we are trying to figure out what's in the mind of another person—that is, his intentions, beliefs, or feelings. The brain region most directly associated with mentalizing is a region of the medial PFC. This mentalizing region is near, though not the same as, the region involved in self-reflection. Thus, mentalizing about another's mind or one's own mind recruits the medial PFC.

What about when we make sense of another person's bodily movements without the intent to understand what's going on in the person's mind? For instance, when we imitate someone's finger-tapping, we need not consider that person's state of mind. In this case, activity is consistently observed in the lateral PFC and lateral PAC. Together, these regions are often referred to as the *mirror-neuron* system, because in other primates, single-cell recordings have shown that whether a primate performs an action (reaching for food, say) or just watches another performing this action, the same neurons in the lateral PFC and lateral PAC respond.[4] As with self-processing, we see a split between processing others in terms of their mind or their body.

In both self- and other-processing, medial activations dominate when one is trying to make sense of the target's mind, and

Nature 419(2002): 469–70; V. Ganesan et al., "Schneiderian First-Rank Symptoms and Right Parietal Hyperactivation: A Replication Using fMRI," *American Journal of Psychiatry* 162(2005): 1545.

[4] M. Iacoboni et al., "Cortical Mechanisms of Human Imitation," *Science* 286(1999): 2526–28; G. Rizzolatti and L. Craighero, "The Mirror-Neuron System," *Annual Review of Neuroscience* 27(2004): 169–92.

lateral activations dominate when one focuses on the target's body. The brain regions are in relatively similar locations (that is, the PFC and PAC) on the medial and lateral surfaces of the brain, but they are quite distinct, based on the focus of attention either on minds or on bodies. Additionally, activation in the lateral regions is associated with reduced activity in the medial regions,[5] suggesting that—at least under some conditions—the activity in the medial and lateral regions may be competitive.

Thus, minds and bodies are represented in the brain in distinct networks, creating a kind of dualism within the brain. Generally speaking, when the brain processes two things in different brain networks, those two things are experienced as being in separate categories. For instance, colors and numbers are experienced as separate categories and are processed in discrete neural networks. (Interestingly, rare individuals called synesthesiacs see colors for numbers or conflate other such "qualia"—for example, "seeing" music or "tasting" visual stimuli. The UCSD neuroscientist V. S. Ramachandran has shown that such people tend to process these separate qualia in the same brain area.) Because of this normal separation in the brain, trying to convince people that minds and bodies are really one kind of thing rather than two might be like trying to convince them that colors and numbers are one kind of thing. It doesn't matter what science tells us, it just isn't borne out by our immediate daily experience.

[5] K. A. McKiernan et al., "A Parametric Manipulation of Factors Affecting Task-Induced Deactivation in Functional Neuroimaging," *Journal of Cognitive Neuroscience* 15(2003): 394–408.

Recall that the second indicator of the Deacon doctrine is that sticky ideas may evolve from less sticky ideas as the ideas transform to better fit the structure and function of the brain. Such idea evolution appears to have occurred with mind-body dualism. Dualism was hardly a new idea when Descartes wrote about it; previous proponents include Pythagoras, Cicero, Saint Augustine, and Thomas Aquinas. The most well-known of these pre-Cartesian dualisms is Plato's; he proposed a theory that contrasted the physical world with the world of universal forms, suggesting that we could appreciate a particular chair as a member of the chair category because we had access to the universal idea of "chair." These universal ideas existed within their own realm, rather than in the mind or the body. While Plato's theory was influential in philosophical circles, it never caught on as a common idea among the masses, and no social policy has ever turned on our feelings about universal forms. Could this be because there are no brain structures devoted to processing universal forms? Are universal forms just one of a countless number of propositional schemes that the all-purpose symbolic machinery of the brain can process but that it need not process—any more than the dorm-room "discovery" that the planets orbiting the sun are analogous to electrons around the atomic nucleus or that the Milky Way is just one molecule in a vast cosmic entity. We can entertain the planet/electron idea, but it isn't sticky, and neither is universal-forms versus physical-world dualism.

A variety of dualisms were proposed to account for many of the same complexities of the world, and yet none really stuck until Descartes' version. This version just happens to correspond to a major division in how the brain processes minds and

bodies. Despite the concerted efforts of scientists and philosophers to discredit mind-body dualism, it remains a core belief
and way of processing the world.

Let us turn to the second Big Idea that the Deacon doctrine
may explain.

Eastern versus Western Culture

Since the early 1990s, there have been fevered debates in psychology over whether and how a particular culture shapes the
minds of those raised in it. The conceptual breakthrough,
which has led to hundreds of studies, came in 1991, from Hazel
Markus of Stanford and Shinobu Kitayama of Kyoto University,
who suggested that Eastern and Western cultures tend to incul-
cate, respectively, interdependent and independent frames for
seeing the world and one's place in it. In essence, East Asians
are raised to believe that we are all connected and that the
needs of the group outweigh the needs of the individual. In contrast, people from Western Europe and North America are
taught to prioritize their own goals, feelings, and achievements.
Social rewards and punishments follow accordingly, such that in
interdependent (Eastern) cultures "the nail that stands out gets
pounded down," whereas in independent (Western) cultures
"the squeaky wheel gets the grease." Being raised in one culture
or the other is thought to shape one's mind such that the world
comes to be seen in interdependent or independent ways, leading individuals to live in accordance with their culture's values.

The values of each culture represent a culture-specific Big
Idea that has endured in each culture for more than a millen-

nium. The standard account is that the cultures shape minds and brains. The Deacon doctrine would suggest that the opposite explanation may also hold true. That is, what if East Asians and Western Europeans have brains that differ in just the right ways, such that each culture's Big Idea would be sticky? What if differences in the brains of people in these geographical regions promote cultural narratives that lead each group to value those ways of organizing society that reflect the group's type of brain organization? For instance, if the people of one culture had congenitally poor hearing and the people of another had congenitally poor sight, they would no doubt value music and art differently.

Baldwin Way, a postdoctoral fellow in my lab at UCLA, has recently come across a key genetic difference between individuals of Eastern and Western descent that differentially affects their brains. A subsequent series of conversations led us to begin testing this idea. Way was reviewing research on genes that control the brain's serotonin system. He discovered that individuals of Eastern and Western descent show differentially distributed variations within the regulatory region of the serotonin transporter gene (5-HTTLPR). There are three different forms of the 5-HTTLPR genetic polymorphism, based on the combination of two alleles; these variants (for shorthand) are called short-short, long-short, and long-long. Whereas two-thirds of East Asians have the short-short variant, only one-fifth of Americans and Western Europeans have it. This is an enormous and highly reliable difference, seen in multiple studies.[6]

[6] See, for example, J. Gelernter et al., "Serotonin Transporter Protein (SLC6A4) Allele and Haplotype Frequencies and Linkage Disequilibria in African- and European-American and Japanese Populations and in Alcohol-Dependent Subjects," *Human Genetics* 101(1997): 243–46.

The serotonin system, and this gene in particular, is related to socioemotional sensitivity. For instance, in one study, children with the short-short variant were shown to be at higher risk for depression, but only if they lacked social support; whereas the risk for depression in those with the long-short and long-long gene variants remained unaffected by social support.[7] Another study found that short-short individuals from nonsupportive families had the greatest depressive symptomology and short-short individuals from supportive families had the least depressive symptomology—with individuals possessing the long-short and long-long gene variants falling in the middle, regardless of whether their family background was supportive or not.[8] These results suggest that the well-being of those with the short-short variant of the 5-HTTLPR gene is more dependent on the quality of the social environment and that these individuals are likely to be more sensitive to the social environment in general.

In light of the Deacon doctrine, the prevalence of short-short 5-HTTLPR polymorphism in individuals of East Asian descent suggests that they may possess the kind of neurochemistry that would predispose them toward interdependence, establishing this as a cultural value, or enduring Big Idea, in this region of the world. If your well-being tends to be dependent on

[7] J. Kaufman et al., "Social Supports and Serotonin Transporter Gene Moderate Depression in Maltreated Children," *Proceedings of the National Academy of Sciences* 101(2004): 17316–21.

[8] S. E. Taylor et al., "Early Family Environment, Current Adversity, the Serotonin Transporter Promoter Polymorphism, and Depressive Symptomatology," *Biological Psychiatry* 60, no. 7(2006): 671–76.

how you are treated by others, then you would certainly prefer a culture that encourages others to make your well-being a priority. In contrast, the relative absence of this gene type in the West would lead to a neurochemistry predisposing people to create a culture that values independence and individual achievement.

Recall that mind-body dualism predates Descartes yet hasn't evolved further since his formulation of it, despite numerous critiques. In the case of Eastern and Western cultural differences, over time there has been, analogous to the evolution of an idea, a territorial migration of the two cultural ideas—both of which seem to have originated in central Asia—with one moving nearly exclusively eastward and the other almost exclusively westward.

Eastern and Western cultures can each reasonably be described as the combination of a religion with a particular brand of civics. Eastern culture solidified in the form of neo-Confucianism, which combined the Buddhist beliefs that we are all connected and that selfish attachments are unhealthy with Confucian civics, which characterizes society in terms of the relational obligations among its members. Western culture emerged out of the combination of Judeo-Christian theology, which posits a single god who holds individuals responsible for their own eternal salvation, and Greek civics, which emphasized personal agency and free will.

Buddhism spread from India toward East Asian countries, with more lasting effects the farther east it traveled; there is no longer a major Buddhist presence in India, where less than half the population has the short-short variant of the 5-HTTLPR gene. In contrast, Christianity spread from the Middle East westward toward Europe and then on to North America, with more lasting effects the farther west it traveled. In both cases, the Big Ideas started out as relatively small ideas that had to travel thousands of miles to find the hospitable regions where they could flourish and become Big Ideas. Interestingly, there were contacts between leaders and representatives of Eastern and Western religions in the time of Alexander the Great, during the Roman Empire, and again in the medieval era. Although the religions were adopted with relative ease when headed in their natural direction—for example, Buddhism to the East— virtually no cross-fertilization of these religions has occurred until lately, when the global economy of the twentieth century eclipsed existing constraints. These cultural Big Ideas appear to have migrated until they found the populations with the right neurochemistry to make them sticky.

We like to think of our beliefs as stemming from some combination of logical analysis and peer influence. The Deacon doctrine suggests another route: the human brain is predisposed to find some ideas appealing because of the structural fit between itself and the idea in question. In the case of Cartesian dualism, we've seen that the brain represents minds and bodies in discrete neural circuits, presumably giving rise to the immediate experience of, and consequent belief in, minds and bodies as discrete categories despite evidence to the contrary. In the case of Eastern and Western cultures, we've seen that regional genetic variation has given rise to distinct brain chemistries that render the two populations differentially sensitive to social feedback and thus differentially receptive to cultural beliefs and values that do or do not prioritize social interdependence. In both cases, the Deacon doctrine uses neuroscience to provide counterintuitive explanations of some of our most deeply held beliefs. When enough brains are predisposed to find the same idea compelling, it is likely to stick around for quite some time.

JOSHUA D. GREENE,

a cognitive neuroscientist and a philosopher, received his bachelor's degree in philosophy from Harvard University (1997) and his Ph.D. from Princeton University (2002). In 2006 he joined the faculty of Harvard's Department of Psychology as an assistant professor. His primary research interest is the psychological and neuroscientific study of morality, focusing on the interplay between emotional and "cognitive" processes in moral decision making. His broader interests cluster around the intersection of philosophy, psychology, and neuroscience. He is currently writing a book about the philosophical implications of our emerging scientific understanding of morality.

FRUIT FLIES OF THE MORAL MIND

JOSHUA D. GREENE

Consider the following moral dilemma: It's wartime. You and your fellow villagers are hiding from nearby enemy soldiers in a basement. Your baby starts to cry, and you cover its mouth to block the sound. If you remove your hand, your baby will cry loudly and the soldiers will hear. They will find you, your baby, and the others, and they will kill all of you. If you do not remove your hand, your baby will suffocate. Is it morally acceptable to smother your baby to death in order to save yourself and the other villagers?

It's unsettling to think about such questions, but it's also instructive. This dilemma, known as the Crying Baby dilemma, nicely captures the tension between two major schools of moral and political thought. On the one hand, we have the utilitarians, philosophers like Jeremy Bentham and John Stuart Mill; according to them, acting morally is ultimately a matter of producing the best overall consequences, striving for "the greater good." On the other hand, we have the deontologists, philosophers like Immanuel Kant, who think that rights and duties often trump the greater good. In the Crying Baby dilemma, the greater good (at least in terms of the number of lives saved) is served by smothering the baby. But many would say that smothering the baby, in addition to being tragic and difficult to do, would also be morally wrong—a violation of the baby's rights, the parent's duty, or both.

The Crying Baby dilemma is also a window into the organization of the human brain. People often speak of a "moral faculty" or a "moral sense," suggesting that moral judgment is a unified phenomenon, but recent advances in the scientific study of moral judgment paint a very different picture. Moral judgment, it seems, depends on a complex interplay between intuitive emotional responses and more effortful "cognitive" processes. More specifically, it seems that intuitive emotional responses to harmful actions ("Don't smother the baby!") depend on one set of brain systems, whereas our more controlled, cognitive responses ("Smothering the baby promotes the greater good") depend on a different set of brain systems. When we puzzle over such moral dilemmas, these neural systems compete, and our all-too-human sense of anguish is the product of that competition. If I'm right, this tension between competing neural systems underlies not only centuries-old disagreements between the likes of Mill and Kant but also contemporary tussles over such issues as stem-cell research and the torturing of suspected terrorists.

Let's consider a pair of moral dilemmas that are a part of a thought experiment known as the Trolley Problem, a staple of contemporary ethics. The first of these we'll call the Switch dilemma, and it goes like this: A runaway trolley is about to run over and kill five people, but you can save them by hitting a switch that will divert the trolley onto a side track, where it will run over and kill only one person. Is it okay to hit the switch? Here, most people say yes, consistent with utilitarian philosophy. Next consider the Footbridge dilemma: Here, too, a runaway trolley threatens five people, but this time instead of standing by a switch you are standing on a footbridge spanning the tracks, in between the oncoming trolley and the unsuspecting five. Next to you is a large man, and the only way to save the five people is to push this large man off the footbridge and into the trolley's path, stopping the trolley but killing your human trolley-stopper in the process. Is it okay to push this man to his death in order to save the five? (I know what you're thinking, and I'll have none of it: No, you can't jump yourself. You're not big enough to stop the trolley. No, you can't shout a warning to the people on the tracks. Yes, the trolley is sure to kill all five people. Yes, your aim will be perfect and the large man will indeed stop the trolley. No, the large man is not Osama bin Laden and the people on the tracks are not your parents, your two children, and your personal trainer. In short, you may not rewrite the question to make it easier.) In response to this case—properly interpreted—most people judge that it would be wrong to sacrifice one life to save five. And here Kant et al. carry the day, as most people place the rights of the man on the footbridge above the greater good.

Why do we go with numbers in the first case but not in the second? Several years ago, I had a hunch that the action in the Footbridge dilemma, with the up-close-and-personal pushing, is more emotionally salient than the action in the Switch dilemma, and that this difference in emotional response could explain why we respond so differently to these two cases. My collaborators and I tested this hypothesis by scanning people's brains while they contemplated dilemmas like the Footbridge dilemma, which we called "personal dilemmas," and dilemmas like the Switch dilemma, which we called "impersonal dilemmas." Our hypothesis predicted that the personal dilemmas would elicit increased activity in parts of the brain associated with emotion, whereas the impersonal dilemmas would elicit increased activity in parts of the brain associated with more effortful, cognitive processes, such as reasoning. And that's what we found.[1] More specifically, responding to personal dilemmas, such as the Footbridge dilemma, elicited increased activity in the medial prefrontal cortex, along with other brain regions associated with emotion and social thinking. Impersonal dilemmas, like the Switch dilemma, by contrast, elicited increased activity in the dorsolateral prefrontal cortex, a classically "cognitive" part of the brain that becomes more active when, say, you're holding in mind a phone number.

What does this tell us about moral thinking? Here's the idea: In response to both the Switch and Footbridge dilemmas, people engage in utilitarian reasoning. ("Five lives at the cost of one? Sounds like a good deal.") But in response to the more personal harm proposed in the Footbridge case, there is also a negative emotional response that says, "No! Don't push that man!" and this response tends to dominate the decision. The emotional response in the Switch dilemma is considerably weaker;

---

[1] J. D. Greene et al., "An fMRI Investigation of Emotional Engagement in Moral Judgment," *Science* 293, no. 5537(2001):2105–8.

thus utilitarian reasoning dominates the decision and we vote for saving the five. The emotional response that dominates the decision in the Footbridge dilemma depends on neural activity in emotion-related brain regions such as the medial prefrontal cortex, whereas the more actuarial thinking that dominates the decision in the Switch dilemma depends on neural activity in classically "cognitive" brain regions such as the dorsolateral prefrontal cortex.

In a follow-up experiment, we focused on dilemmas of greater difficulty, such as the Crying Baby case. These, too, are personal dilemmas, but they're constructed so that the utilitarian rationale is stronger. In the Footbridge case, it's one life versus five, but in the Crying Baby case *everyone* dies if you don't act, including you and your baby. In response to the Switch and Footbridge dilemmas, people's judgments are fairly consistent, but in response to the Crying Baby dilemma, people's judgments are split about fifty-fifty and nearly everyone takes a long time to respond.

What's going on? If the theory I've described is correct, the Crying Baby dilemma triggers a conflict between the emotional and cognitive parts of the brain. Conveniently, there is a part of the brain called the anterior cingulate cortex that reliably responds to this kind of internal conflict. When your brain is trying to do two different things at once, the anterior cingulate cortex says, "Houston, we have a problem." We predicted that this area would become more active in response to dilemmas like the Crying Baby case, and indeed it does.

If the anterior cingulate cortex says, "Houston, we have a problem," this naturally raises the question: where's Houston? Houston, it turns out, is in the dorsolateral prefrontal cortex, where we keep phone numbers in mind and engage in abstract reasoning. This brain area also gives us our ability to resist impulses. The common theme among these operations is *cognitive control*—the ability to guide attention, thought, and action in accordance with goals or intentions. If the thought of harming someone in a "personal" way triggers an emotional response that makes us say "No!" then approving of a "personal" harm that promotes the greater good requires the ability to override that emotional response. And *that* requires increased activity in the dorsolateral prefrontal cortex, the seat of cognitive control. This suggests that when people make utilitarian judgments in response to difficult dilemmas like the Crying Baby case, they should exhibit increased activity in their dorsolateral prefrontal cortices, which is what we found.[2] A more recent study of ours fits the same pattern. We had people consider dilemmas in which promoting the greater good required breaking a promise, and, as before, we saw more activity in the dorsolateral prefrontal cortex when people gave utilitarian answers favoring the greater good.

This dual-process theory of moral judgment—"dual process" because it posits distinct emotional and cognitive processes—makes some interesting predictions about the behavior of neurological patients. For example, patients with frontotemporal dementia (FTD) are known for their "emotional blunting." A team from UCLA presented FTD patients with versions of the Switch and Footbridge dilemmas. Their responses to the Switch dilemma were pretty standard, but

[2] J. D. Greene et al., "The Neural Bases of Cognitive Conflict and Control in Moral Judgment," *Neuron* 44, no. 2(2004): 389–400.

they were far more likely than others to approve of pushing the man off the footbridge. Without the emotions to tell them "No!" this action, too, seemed like a "good deal." Two other research teams, one in Iowa and one in Italy, got similar results testing patients with damage to the ventromedial prefrontal cortex, a region known to be important for emotion-based decision making. Both groups found that these patients gave unusually utilitarian responses to dilemmas like the Footbridge and Crying Baby cases. The Iowa patients, in fact, were almost five times more likely than control subjects to give utilitarian responses.[3]

Piercarlo Valdesolo and David DeSteno at Northeastern University used a clever, low-tech trick to make the same point. They presented people with versions of the Switch and Footbridge dilemmas under two different conditions. At the start of the experiment, some people watched a funny film clip from *Saturday Night Live*, while others watched a clip with no particular emotional content. People's responses to the Switch dilemma were unaffected by the choice of film, but the ones who watched the funny *SNL* clip were almost four times more likely to approve of pushing the man off the footbridge. The idea here is that a dose of positive emotion can neutralize the negative emotion that would otherwise make people uncomfortable with pushing the man off the footbridge.

[3] E. Ciaramelli et al., "Selective Deficit in Personal Moral Judgment Following Damage to Ventromedial Prefrontal Cortex," *Social Cognitive and Affective Neuroscience* 2, no. 2(2007): 84–92; M. Koenigs et al., "Damage to the Prefrontal Cortex Increases Utilitarian Moral Judgments," *Nature* 446(7138): 908–11 (2007); M. F. Mendez et al., "An Investigation of Moral Judgment in Frontotemporal Dementia," *Cognitive and Behavioral Neurology* 18, no. 4(2005): 193–97.

My colleagues and I conducted a similar experiment, targeting cognitive control processes rather than emotional ones. In our experiment, people had to make their judgments while simultaneously keeping an eye on a stream of numbers scrolling across a computer screen. Every time the number five went by, they had to hit a button. This kind of nuisance task is known as a "cognitive load," the purpose of which is to gum up the sorts of higher-level cognitive processes that are based in the dorsolateral prefrontal cortex. We found that the cognitive load made people slower to give utilitarian answers ("Smother the baby in the name of the greater good") but had no effect on the characteristically deontological answers ("Don't smother the baby, even if everyone will die"). (In fact, the cognitive load seemed to speed up the deontological answers, but this effect was not statistically significant for the group as a whole.) These two studies are like mirror images: Block the emotional processes and utilitarian judgments come more easily. Block the controlled cognitive processes, and utilitarian judgments come more slowly.

These results are part of a general pattern, one that philosophers may find surprising. When an apparent moral duty ("Don't use people as trolley stoppers") conflicts with the greater good ("Better to save five lives"), judgments in favor of duty are driven by emotion, while judgments in favor of the greater good are driven by more controlled cognitive processes. This is surprising because philosophers like Kant, who place duty above the greater good, are often regarded as rationalists, philosophers whose moral conclusions are supposed to be grounded in reason. But the studies described here suggest that this kind of philosophy is less about reasoning and more about *rationalizing*. My colleague Jonathan Haidt thinks that pretty much all of moral reasoning is like this, but I disagree.[4] Based on the research described here, I believe that utilitarian judgments are driven by reasoning processes, enabled by the dorsolateral prefrontal cortex, a.k.a. Houston. Of course, I don't think it's all so neat and simple. Following David Hume, I suspect that even utilitarian calculation requires a kind of emotion—less like alarm bells going off and more like sand accumulating on a scale—and there are indeed hints of this in the brain-imaging data.

People sometimes ask me why I bother with these bizarre hypothetical dilemmas. Shouldn't we be studying *real* moral decision making instead? To me, these dilemmas are like a geneticist's fruit flies. They're manageable enough to play around with in the lab but complex enough to capture something interesting about the wider and wilder world outside. With that in mind, let me introduce a final pair of moral dilemmas, originally devised by the utilitarian philosopher Peter Singer.[5]

You're walking by a pond one day, when you spot a small child drowning. You could easily wade in and save her, but this would ruin your stylish new Italian suit. So you walk on by. Are you a terrible person? Yes, we say. Next case: You receive a letter from a reputable international aid organization such as UNICEF or Oxfam. They would like you to donate five hundred dollars, which they will use to save the lives of several poor

[4] J. Haidt, "The Emotional Dog and Its Rational Tail: A Social Intuitionist Approach to Moral Judgment," *Psycholigcal Review* 108(2001): 814–34.
[5] P. Singer, "Famine, Affluence and Morality," *Philosophy and Public Affairs* 1(1972): 229–43.

African children in desperate need of food and medicine. You feel sorry for these children, but you've had your eye on a stylish new Italian suit, and you'd prefer to save your money for that. You toss the letter in the trash. Are you a horrible person? You're no saint, we say, but you certainly haven't done anything wrong.

What's the difference between refusing to save a child who's drowning right in front of you and refusing to save a child who's drowning in poverty on the other side of the world? Your rationalizing mind is already at work on the problem: In the case of the drowning child, you're the only one who can help, but many others can help those poor African children. They're the world's problem, not yours. Fair enough. But what if you're standing around the pond watching the child drown with dozens of other people, all of whom are rather fond of *their* stylish Italian suits? (It's the annual meeting of the American Bar Association.) Now is it okay to let the child drown? We can play this game all day, but we are unlikely to find a satisfying resolution. An alternative approach is to think about the relevant psychology and its natural history.

Let's try this first with the Switch and Footbridge dilemmas. As I've explained, pushing someone to his death is more emotionally salient than hitting a switch that achieves the same thing. But why? An evolutionary perspective may be useful. We evolved in an environment in which good old-fashioned pushing and shoving were prevalent; we did not evolve in an age of mechanically mediated threats. It makes sense, then, that these more basic forms of "personal" violence push our moral buttons, whereas distinctively modern forms of violence do not. Something similar may be true for acts of altruism and the emotions that support them. We did not evolve in an environment in which one could save the lives of distant strangers by being less fashionable, but we did evolve in a world in which one could help desperate people in the here and now. Nature endowed us with tuggable heartstrings, a crucial design feature for creatures whose survival depends on cooperation. But nature couldn't foresee that our survival might someday depend on cooperation across oceans and continents, and so neglected to outfit us with heartstrings that are readily tugged from a distance.

We are, of course, a very clever species. Through our ingenuity, we've made ourselves faster and stronger and more dangerous than all the other creatures on Earth. Perhaps by applying our capacity for complex cognition to the problems of modern life, we can transcend the limitations of our moral instincts.

LERA BORODITSKY

is an assistant professor of psychology, neuroscience, and symbolic systems at Stanford University. Dr. Boroditsky grew up in Minsk in the former Soviet Union. After earning a Ph.D. in cognitive psychology from Stanford in 2001, she served on the faculty at MIT in the Department of Brain and Cognitive Sciences before returning to a faculty position at Stanford. She also runs a satellite laboratory in Jakarta, Indonesia.

Boroditsky's research centers on the nature of mental representation and how knowledge emerges out of the interactions of mind, world, and language. One focus has been to investigate the ways that languages and cultures shape human thinking. To this end, Boroditsky's laboratory has collected data around the world, from Indonesia to Chile to Turkey to Aboriginal Australia. Her research has been widely featured in the media and has won multiple awards, including the CAREER award from the National Science Foundation and the Searle Scholars award.

HOW DOES OUR LANGUAGE SHAPE THE WAY WE THINK?

LERA BORODITSKY

Humans communicate with one another using a dazzling array of languages, each differing from the next in innumerable ways. Do the languages we speak shape the way we see the world, the way we think, and the way we live our lives? Do people who speak different languages think differently simply because they speak different languages? Does learning new languages change the way you think? Do polyglots think differently when speaking different languages?

These questions touch on nearly all of the major controversies in the study of mind. They have engaged scores of philosophers, anthropologists, linguists, and psychologists, and they have important implications for politics, law, and religion. Yet despite nearly constant attention and debate, very little empirical work was done on these questions until recently. For a long time, the idea that language might shape thought was considered at best untestable and more often simply wrong. Research in my labs at Stanford University and at MIT has helped reopen this question. We have collected data around the world: from China, Greece, Chile, Indonesia, Russia, and Aboriginal Australia. What we have learned is that people who speak different languages do indeed think differently and that even flukes of grammar can profoundly affect how we see the world. Language is a uniquely human gift, central to our experience of being human. Appreciating its role in constructing our mental lives brings us one step closer to understanding the very nature of humanity.

I often start my undergraduate lectures by asking students the following question: which cognitive faculty would you most hate to lose? Most of them pick the sense of sight; a few pick hearing. Once in a while, a wisecracking student might pick her sense of humor or her fashion sense. Almost never do any of them spontaneously say that the faculty they'd most hate to lose is language. Yet if you lose (or are born without) your sight or hearing, you can still have a wonderfully rich social existence. You can have friends, you can get an education, you can hold a job, you can start a family. But what would your life be like if you had never learned a language? Could you still have friends, get an education, hold a job, start a family? Language is so fundamental to our experience, so deeply a part of being human, that it's hard to imagine life without it. But are languages merely tools for expressing our thoughts, or do they actually shape our thoughts?

Most questions of whether and how language shapes thought start with the simple observation that languages differ from one another. And a lot! Let's take a (very) hypothetical example. Suppose you want to say, "Bush read Chomsky's latest book." Let's focus on just the verb, "read." To say this sentence in English, we have to mark the verb for tense; in this case, we have to pronounce it like "red" and not like "reed." In Indonesian you need not (in fact, you can't) alter the verb to mark tense. In Russian you would have to alter the verb to indicate tense and gender. So if it was Laura Bush who did the reading, you'd use a different form of the verb than if it was George. In Russian you'd also have to include in the verb information about completion. If George read only part of the book, you'd use a different form of the verb than if he'd diligently plowed through the whole thing. In Turkish you'd have to include in the verb how you acquired this information: if you had witnessed this unlikely event with your own two eyes, you'd use one verb form, but if you had simply read or heard about it, or inferred it from something Bush said, you'd use a different verb form.

Clearly, languages require different things of their speakers. Does this mean that the speakers think differently about the world? Do English, Indonesian, Russian, and Turkish speakers end up attending to, partitioning, and remembering their experiences differently just because they speak different languages?

For some scholars, the answer to these questions has been an obvious yes. Just look at the way people talk, they might say. Certainly, speakers of different languages must attend to and encode strikingly different aspects of the world just so they can use their language properly.

Scholars on the other side of the debate don't find the differences in how people talk convincing. All our linguistic utterances are sparse, encoding only a small part of the information we have available. Just because English speakers don't include the same information in their verbs that Russian and Turkish speakers do doesn't mean that English speakers aren't paying attention to the same things; all it means is that they're not talking about them. It's possible that everyone thinks the same way, notices the same things, but just talks differently.

Believers in cross-linguistic differences counter that everyone does *not* pay attention to the same things: if everyone did, one might think it would be easy to learn to speak other languages. Unfortunately, learning a new language (especially one not closely related to those you know) is never easy; it seems to require paying attention to a new set of distinctions. Whether it's distinguishing modes of being in Spanish, evidentiality in Turkish, or aspect in Russian, learning to speak these languages requires something more than just learning vocabulary: it requires paying attention to the right things in the world so that you have the correct information to include in what you *say*.

Such a priori arguments about whether or not language shapes thought have gone in circles for centuries, with some arguing that it's impossible for language to shape thought and others arguing that it's impossible for language *not* to shape thought. Recently my group and others have figured out ways to empirically test some of the key questions in this ancient debate, with fascinating results. So instead of arguing about what must be true or what can't be true, let's find out what *is* true.

Follow me to Pormpuraaw, a small Aboriginal community on the western edge of Cape York, in northern Australia. I came here because of the way the locals, the Kuuk Thaayorre, talk about space. Instead of words like "right," "left," "forward," and "back," which, as commonly used in English, define space relative to an observer, the Kuuk Thaayorre, like many other Aboriginal groups, use cardinal-direction terms—north, south, east, and west—to define space.[1] This is done at all scales, which means you have to say things like "There's an ant on your southeast leg" or "Move the cup to the north-northwest a little bit." One obvious consequence of speaking such a language is that you have to stay oriented at all times, or else you cannot speak properly. The normal greeting in Kuuk Thaayorre is "Where are you going?" and the answer should be something like "South-southeast, in the middle distance." If you don't know which way you're facing, you can't even get past "Hello."

The result is a profound difference in navigational ability and spatial knowledge between speakers of languages that rely primarily on absolute reference frames (like Kuuk Thaayorre) and languages that rely on relative reference frames (like English).[2] Simply put, speakers of languages like Kuuk Thaayorre are much better than English speakers at staying oriented and keeping track of where they are, even in unfamiliar landscapes or inside unfamiliar buildings. What enables them—in fact, forces them—to do this is their language. Having their attention trained in this way equips them to perform navigational feats once thought beyond human capabilities.

Because space is such a fundamental domain of thought, differences in how people think about space don't end there. People rely on their spatial knowledge to build other, more complex, more abstract representations. Representations of such things as time, number, musical pitch, kinship relations,

[1] S. C. Levinson and D. P. Wilkins, eds., *Grammars of Space: Explorations in Cognitive Diversity* (New York: Cambridge University Press, 2006).
[2] Levinson, *Space in Language and Cognition: Explorations in Cognitive Diversity* (New York: Cambridge University Press, 2003).

morality, and emotions have been shown to depend on how we think about space. So if the Kuuk Thaayorre think differently about space, do they also think differently about other things, like time? This is what my collaborator Alice Gaby and I came to Pormpuraaw to find out.

To test this idea, we gave people sets of pictures that showed some kind of temporal progression (e.g., pictures of a man aging, or a crocodile growing, or a banana being eaten). Their job was to arrange the shuffled photos on the ground to show the correct temporal order. We tested each person in two separate sittings, each time facing in a different cardinal direction. If you ask English speakers to do this, they'll arrange the cards so that time proceeds from left to right. Hebrew speakers will tend to lay out the cards from right to left, showing that writing direction in a language plays a role.[3] So what about folks like the Kuuk Thaayorre, who don't use words like "left" and "right"? What will they do?

The Kuuk Thaayorre did not arrange the cards more often from left to right than from right to left, nor more toward or away from the body. But their arrangements were not random: there was a pattern, just a different one from that of English speakers. Instead of arranging time from left to right, they arranged it from east to west. That is, when they were seated facing south, the cards went left to right. When they faced north, the cards went from right to left. When they faced east,

---

[3] B. Tversky et al., "Cross-Cultural and Developmental Trends in Graphic Productions," *Cognitive Psychology* 23(1991): 515–57; O. Fuhrman and L. Boroditsky, "Mental Time-Lines Follow Writing Direction: Comparing English and Hebrew Speakers," *Proceedings of the 29th Annual Conference of the Cognitive Science Society* (2007): 1007–10.

the cards came toward the body, and so on. This was true even though we never told any of our subjects which direction they faced. The Kuuk Thaayorre not only knew that already (usually much better than I did), but they also spontaneously used this spatial orientation to construct their representations of time.

People's ideas of time differ across languages in other ways. For example, English speakers tend to talk about time using horizontal spatial metaphors (e.g., "The best is *ahead* of us," "The worst is *behind* us"), whereas Mandarin speakers have a vertical metaphor for time (e.g., the next month is the "down month" and the last month is the "up month"). Mandarin speakers talk about time vertically more often than English speakers do, so do Mandarin speakers think about time vertically more often than English speakers do? Imagine this simple experiment. I stand next to you, point to a spot in space directly in front of you, and tell you, "This spot, here, is today. Where would you put yesterday? And where would you put tomorrow?" When English speakers are asked to do this, they nearly always point horizontally. But Mandarin speakers often point vertically, about seven or eight times more often than do English speakers.[4]

Even basic aspects of time perception can be affected by language. For example, English speakers prefer to talk about duration in terms of length (e.g., "That was a *short* talk," "The meeting didn't take *long*"), while Spanish and Greek speakers prefer to talk about time in terms of amount, relying more on

[4] L. Boroditsky, "Do English and Mandarin Speakers Think Differently About Time?" *Proceedings of the 48th Annual Meeting of the Psychonomic Society* (2007): 34.

words like "much," "big," and "little" rather than "short" and "long." Our research into such basic cognitive abilities as estimating duration shows that speakers of different languages differ in ways predicted by the patterns of metaphors in their language. (For example, when asked to estimate duration, English speakers are more likely to be confused by distance information, estimating that a line of greater length remains on the test screen for a longer period of time, whereas Greek speakers are more likely to be confused by amount, estimating that a container that is fuller remains longer on the screen.)[5]

An important question at this point is: are these differences caused by language per se or by some other aspect of culture? Of course, the lives of English, Mandarin, Greek, Spanish, and Kuuk Thaayorre speakers differ in a myriad of ways. How do we know that it is language itself that creates these differences in thought and not some other aspect of their respective cultures?

One way to answer this question is to teach people new ways of talking and see if that changes the way they think. In our lab, we've taught English speakers different ways of talking about time. In one such study, English speakers were taught to use size metaphors (as in Greek) to describe duration (e.g., a movie *is larger than* a sneeze), or vertical metaphors (as in Mandarin) to describe event order. Once the English speakers had learned to talk about time in these new ways, their cognitive performance began to resemble that of Greek or Mandarin

---

[5] D. Casasanto et al., "How Deep Are Effects of Language on Thought? Time Estimation in Speakers of English, Indonesian, Greek, and Spanish," *Proceedings of the 26th Annual Conference of the Cognitive Science Society* (2004): 575–80.

speakers. This suggests that patterns in a language can indeed play a causal role in constructing how we think.[6] In practical terms, it means that when you're learning a new language, you're not simply learning a new way of talking, you are also inadvertently learning a new way of thinking.

Beyond abstract or complex domains of thought like space and time, languages also meddle in basic aspects of visual perception—our ability to distinguish colors, for example. Different languages divide up the color continuum differently: some make many more distinctions between colors than others, and the boundaries often don't line up across languages.

To test whether differences in color language lead to differences in color perception, we compared Russian and English speakers' ability to discriminate shades of blue. In Russian there is no single word that covers all the colors that English speakers call "blue." Russian makes an obligatory distinction between light blue (*goluboy*) and dark blue (*siniy*). Does this distinction mean that *siniy* blues look more different from *goluboy* blues to Russian speakers? Indeed, the data say yes. Russian speakers are quicker to distinguish two shades of blue that are called by the different names in Russian (i.e., one being *siniy* and the other being *goluboy*) than if the two fall into the same category. For English speakers, all these shades are still designated by the same word, "blue," and there are no comparable differences in reaction time.

Further, the Russian advantage disappears when subjects

---

[6] Ibid., "How Deep Are Effects of Language on Thought? Time Estimation in Speakers of English and Greek" (in review); L. Boroditsky, "Does Language Shape Thought? English and Mandarin Speakers' Conceptions of Time," *Cognitive Psychology* 43, no. 1(2001): 1–22.

are asked to perform a verbal interference task (reciting a string of digits) while making color judgments but not when they're asked to perform an equally difficult spatial interference task (keeping a novel visual pattern in memory). The disappearance of the advantage when performing a verbal task shows that language is normally involved in even surprisingly basic perceptual judgments—and that it is language per se that creates this difference in perception between Russian and English speakers. When Russian speakers are blocked from their normal access to language by a verbal interference task, the differences between Russian and English speakers disappear.

Even what might be deemed frivolous aspects of language can have far-reaching subconscious effects on how we see the world. Take grammatical gender. In Spanish and other Romance languages, nouns are either masculine or feminine. In many other languages, nouns are divided into many more genders ("gender" in this context meaning class or kind). For example, some Australian Aboriginal languages have up to sixteen genders, including classes of hunting weapons, canines, things that are shiny, or, in the phrase made famous by cognitive linguist George Lakoff, "women, fire, and dangerous things."

What it means for a language to have grammatical gender is that words belonging to different genders get treated differently grammatically and words belonging to the same grammatical gender get treated the same grammatically. Languages can require speakers to change pronouns, adjective and verb endings, possessives, numerals, and so on, depending on the noun's gender. For example, to say something like "my chair was old" in Russian (*moy stul bil' stariy*), you'd need to make every word in the sentence agree in gender with "chair" (*stul*), which is masculine in Russian. So you'd use the masculine form of "my," "was," and "old." These are the same forms you'd use in speaking of a biological male, as in "my grandfather was old." If, instead of speaking of a chair, you were speaking of a bed (*krovat*), which is feminine in Russian, or about your grandmother, you would use the feminine form of "my," "was," and "old."

Does treating chairs as masculine and beds as feminine in the grammar make Russian speakers think of chairs as being more like men and beds as more like women in some way? It turns out that it does. In one study, we asked German and Spanish speakers to describe objects having opposite gender assignment in those two languages. The descriptions they gave differed in a way predicted by grammatical gender. For example, when asked to describe a "key"—a word that is masculine in German and feminine in Spanish—the German speakers were more likely to use words like "hard," "heavy," "jagged," "metal," "serrated," and "useful," whereas Spanish speakers were more likely to say "golden," "intricate," "little," "lovely," "shiny," and "tiny." To describe a "bridge," which is feminine in German and masculine in Spanish, the German speakers said "beautiful," "elegant," "fragile," "peaceful," "pretty," and "slender," and the Spanish speakers said "big," "dangerous," "long," "strong," "sturdy," and "towering." This was true even though all testing was done in English, a language without grammatical gender. The same pattern of results also emerged in entirely nonlinguistic tasks (e.g., rating similarity between pictures). And we can also show that it is aspects of language per se that shape how people think: teaching English speakers new grammatical gender systems influences mental representations of objects in the same way it does with German and Spanish speakers. Apparently even small flukes of grammar, like the seemingly arbitrary assignment of gender to a noun, can have an effect on people's ideas of concrete objects in the world.[7]

In fact, you don't even need to go into the lab to see these effects of language; you can see them with your own eyes in an art gallery. Look at some famous examples of personification in art—the ways in which abstract entities such as death, sin, victory, or time are given human form. How does an artist decide whether death, say, or time should be painted as a man or a woman? It turns out that in 85 percent of such personifications, whether a male or female figure is chosen is predicted by the grammatical gender of the word in the artist's native language. So, for example, German painters are more likely to paint death as a man, whereas Russian painters are more likely to paint death as a woman.

The fact that even quirks of grammar, such as grammatical gender, can affect our thinking is profound. Such quirks are pervasive in language; gender, for example, applies to all nouns, which means that it is affecting how people think about anything that can be designated by a noun. That's a lot of stuff!

I have described how languages shape the way we think about space, time, colors, and objects. Other studies have found effects of language on how people construe events, reason about causality, keep track of number, understand material substance, perceive and experience emotion, reason about other people's minds, choose to take risks, and even in the way

[7] L. Boroditsky et al., "Sex, Syntax, and Semantics," in D. Gentner and S. Goldin-Meadow, eds., *Language in Mind: Advances in the Study of Language and Cognition* (Cambridge, MA: MIT Press, 2003), 61–79.

they choose professions and spouses.[8] Taken together, these results show that linguistic processes are pervasive in most fundamental domains of thought, unconsciously shaping us from the nuts and bolts of cognition and perception to our loftiest abstract notions and major life decisions. Language is central to our experience of being human, and the languages we speak profoundly shape the way we think, the way we see the world, the way we live our lives.

[8] L. Boroditsky, "Linguistic Relativity," in L. Nadel, ed., *Encyclopedia of Cognitive Science* (London: MacMillan, 2003), 917–21; B. W. Pelham et al., "Why Susie Sells Seashells by the Seashore: Implicit Egotism and Major Life Decisions," *Journal of Personality and Social Psychology* 82, no. 4(2002): 469–86; A. Tversky and D. Kahneman, "The Framing of Decisions and the Psychology of Choice," *Science* 211(1981): 453–58; P. Pica et al., "Exact and Approximate Arithmetic in an Amazonian Indigene Group," *Science* 306(2004): 499–503; J. G. de Villiers and P. A. de Villiers, "Linguistic Determinism and False Belief," in P. Mitchell and K. Riggs, eds., *Children's Reasoning and the Mind* (Hove, UK: Psychology Press, in press); J. A. Lucy and S. Gaskins, "Interaction of Language Type and Referent Type in the Development of Nonverbal Classification Preferences," in Gentner and Goldin-Meadow, 465–92; L. F. Barrett et al., "Language as a Context for Emotion Perception," *Trends in Cognitive Sciences* 11(2007): 327–32.

SAM COOKE

is a neuroscientist whose work addresses two questions: how do our brains change as we learn, and how do they stay changed, to store memories for a lifetime? He first became interested in these questions while studying philosophy and experimental psychology at the University of Sheffield, United Kingdom, and he received his Ph.D. from University College London in 2002 with a dissertation titled "The Making of a Motor Memory." During this period, he was able to elucidate some of the neural mechanisms underlying the acquisition of simple motor skills through associative learning. He then moved to the National Institute for Medical Research, in North London, where he shifted his focus to the neural basis of episodic memory—that is, memories of the details of the events that one experiences. He also began work on the phenomenon of long-term potentiation, a model of how memories may be stored by modification of synapses in the brain. He is currently a postdoctoral associate at MIT, where he continues to probe the biology of memory.

MEMORY ENHANCEMENT, MEMORY ERASURE

The Future of Our Past

SAM COOKE

In the course of the twenty-first century, there is likely to be a profound shift in the quality of internal human experience. Once we come to understand how our memories are formed, stored, and recalled within the brain, we may be able to manipulate them—to shape our own stories. Our past—or at least our recollection of our past—may become a matter of choice.

We have already seen the arrival of "smart" drugs called nootropics, which increase our learning speed and perhaps our memory capacity. We may develop the technology to select the memories we wish to keep and those we wish to discard. Finally, in the distant future, we may well be able to create memories without having done the learning or lived through the actual experiences. What are the potential ethical ramifications of introducing these technologies into society?

Of all the qualities we possess, it is our memory that best defines us as individuals. Identical twins, for example, have the same genes and look indistinguishable to others, even to those who know them well, but their memories are unique. Memory also allows us to function in the world, providing not just the rose-tinted retrospective of our youth but our understanding of what is right and what is wrong; how we walk, talk, tie our shoelaces; the map we use to navigate in our environment. Memory enables us to tell friend from enemy, family from stranger, fact from fiction. Without memory, we are not just lost; from our own perspective we do not even exist.

Nootropics and Memory Enhancement

The degeneration of memory with age is a defining feature of our time. The average age of humans, particularly in the developed world, is rising at a dramatic rate because of an increase in life expectancy and a reduction in the birth rate. A February 2003 report from the Centers for Disease Control titled "Public Health and Aging" revealed that, worldwide, the number of individuals sixty-five and older is likely to more than double in the 2000–2030 period from 420 million to 973 million, while the world's population increases at a considerably slower rate. Dementia afflicts a large proportion of the elderly; in this country, Alzheimer's disease alone affects 10 percent of those sixty-five or older, and the risk doubles for every five additional years. The loss of identity that occurs and its effect on the family of the sufferer are not the only costs: the economic burden that such mental deterioration places on society is huge and will become greater as the population grows. Much current neuroscientific research is aimed at identifying treatments for age-related memory disorders.

Drugs for the treatment of dementia fall into two groups: those that directly attack deficits resulting from the disorder and those that mask them by enhancing other, unrelated memory processes. This second class of drugs, known as nootropics—a name derived from the ancient Greek *nous* (mind) and *tropein* (to turn)—should also be able to enhance the memory of individuals who do not have dementia. While direct cures for memory disorders are regarded as valuable, research into nootropics is an area of particular interest to pharmaceutical companies because of the much larger market potential. Piracetam was the first such substance to be sold, under the trade name Nootropil. Despite a lack of strong evidence as to its efficacy and mechanism of action, it has proved popular. Potential users range from students studying for final exams to professionals trying to get ahead in the workplace; all are likely to want to enhance their memory and their cognitive powers.

Nootropics themselves can be subdivided into two classes: (1) those that interact directly with the processes of information storage and learning and (2) those that enhance or mimic other important features of the brain that facilitate learning and memory, such as attention, sleep, and reward. Many drugs of the second type have long been with us (nicotine, caffeine, glucose, amphetamines, cocaine), and natural plant and fungal extracts such as gingko biloba, ginseng, and Hydergine (ergoloid mesylates) are already being marketed explicitly as nootropics. Some drugs have been serendipitously identified as nootropics after extensive use for the treatment of other disorders of the mind; two examples are modafinil, a treatment for narcolepsy, and Ritalin, for attention deficit hyperactivity disorder (ADHD). Both of those drugs are now taken recreationally by individuals who do not have either disorder but seek enhanced cognition. The real boon for pharmaceutical companies may well be that other class of nootropics—the substances that tap directly into the molecular processes of learning. It is hoped that they will have fewer unwanted side effects than substances that reduce sleep or enhance attention.

We are still some way from fully unlocking the secrets of memory formation. The high speed with which the brain processes information and its ability to change instantaneously to incorporate new information make it one of the great wonders of nature. Processing speed is achieved by the combined use of electricity (to send impulses along the length of our neurons) and chemicals known as neurotransmitters, which pass information from one neuron to the next at specialized communication points known as synapses. Specifically, at a synapse the neurotransmitters drift across a small gap called the synaptic cleft and are collected on the other side by specialized docking stations called receptors, which convert their chemical energy back into electrical impulses. This process of synaptic transmission is a major focus of scientific investigation.

There are several ways of changing the communication between neurons. The number of neurotransmitter molecules released in an electrical event may be altered, or there may be a change in the number or efficiency of the receptor molecules that collect these neurotransmitters. In addition, alterations in the shape or number of synapses will have a long-term effect on synaptic transmission. These modifications are collectively described as *synaptic plasticity*. Neuroscientists believe that memories may be formed through synaptic plasticity by returning lines of communication in the brain. Pharmaceutical companies have an interest in these processes, too, because they present potential targets for nootropics. Despite much effort, so far there has not been a drug on the market that directly enhances synaptic plasticity. But it is only a matter of time before this happens; such a drug, if it proves safe and effective, will doubtless enjoy huge worldwide sales.

Selective Amnesia

Post-traumatic stress disorder (PTSD) is a condition of memory that lies at the other end of the spectrum from Alzheimer's disease. Here, the brain stores information all too well, scorching the experience of a particularly traumatic event into the minds of sufferers and rendering them crippled by fear. It is a phenomenon that commonly occurs in victims of rape and in soldiers returning from war, and it is the focus of another major thread of current medical research. An effective treatment for PTSD would erase the cause of the fear but spare the cherished memories that make up the human experience. But how can we selectively target one memory without affecting others, given that they all rely on the same biological mechanisms for storage?

It has long been known that there are amnestic agents—substances that, when taken during or even within a short period after learning, prevent a long-lasting memory from forming. Most of us have probably experienced memory loss after an evening in which we drank too much alcohol. Users of cannabis will also be aware that this drug has a negative effect on their memory. Such palliatives are not suitable for PTSD, however, because the disorder arises from unexpected events, which we cannot prepare ourselves for by getting inebriated beforehand. What is needed is a treatment that can be applied after the event, using drugs that affect the process of memory consolidation. Memory consolidation takes place in a limited period after learning, during which the important transition of a memory from an early, fragile state to a later, permanent state occurs. Certain key biological events occur within this consolidation period that stabilize the neural changes underlying memory, and it is these key events that allow us to retain memories for a lifetime.

One important factor appears to be the synthesis of new proteins. It has been known for nearly fifty years that antibiotics, which kill bacteria by preventing them from building new proteins, are amnestic at low doses in a variety of animals, ranging from goldfish to humans.[1] These findings were the first scientific evidence to suggest that our memories are built from new proteins. Temporary changes, hypothetically at the synapse, result from alteration of existing proteins, but lasting changes require their replacement with new proteins. Using protein-synthesis inhibitors to treat PTSD is, however, unlikely to be a viable approach; almost all major cell functions, from division to repair, rely on protein synthesis, so administering inhibitors would have damaging side effects.

An alternative approach targets the systems in the brain that mediate emotional stress responses. The relevant sub-

---

[1] H. P. Davis and L. R. Squire, "Protein Synthesis and Memory: A Review," *Psychological Bulletin* 96, no. 3(1984): 518–59.

stances include adrenaline (also known as epinephrine) and the closely related noradrenaline (norepinephrine), a major neurotransmitter released in large quantities after stressful experiences. During the critical postlearning consolidation period, blocking the receptors that bind noradrenaline using betablockers (agents already prescribed to treat cardiac disorders and stage fright) has been shown to prevent long-term storage of emotional memories in both rats and humans.[2] Propranolol, the active ingredient in beta-blockers, is now being tested in extensive clinical trials as a treatment for PTSD.

This approach has its limits. The propranolol must be given just an hour or so after the precipitating stressful event. In the case of soldiers in combat, one can envisage a scenario in which pills are carried as a standard-issue item and taken immediately after an event that may cause PTSD. In other, more unexpected cases—rape, say, or a car crash—such a treatment may not be readily available; however, recent research into emotional memory suggests a way around this problem. Some forms of memory do not follow the simple, unidirectional timeline from fragile to stable; in fact, certain emotional memories reenter a period in which they are susceptible to erasure by propranolol or protein-synthesis inhibitors whenever they are recalled, due to a phenomenon called reconsolidation.[3] Thus, debilitating memories could be treated at any time, simply by having the

[2] J. Debiec and J. E. LeDoux, "Disruption of Reconsolidation but Not Consolidation of Auditory Fear Conditioning by Noradrenergic Blockade in the Amygdala," *Neuroscience* 129, no. 2(2004): 267–72; J. Giles, "Beta-Blockers Tackle Memories of Horror," *Nature* 436(2005): 448–49.

[3] Debiec and LeDoux, "Noradrenergic Signaling in the Amygdala Contributes to the Reconsolidation of Fear Memory: Treatment Implications for PTSD," *Annals of the New York Academy of Sciences* 1071(2006): 521–24.

sufferer recall the events to spark a state of reconsolidation before taking a dose of propranolol. This approach has not been fully tested in humans, and it may be some time until such a PTSD treatment is used. Nonetheless, it may turn out to have widespread application in this and other memory-related disorders, such as phobias or obsessive-compulsive disorder.

This form of "recall" therapy may well prove of interest to people who do not suffer from such disorders. As with the enhancement of normal human memory using nootropics, memory-erasure technology could have profound social implications—a scenario explored in the film *Eternal Sunshine of the Spotless Mind*, in which unwanted memories of failed love affairs are erased so that people can maintain an idealized view of the world. Could it really become the norm in our future to wipe out all trace of an ex-lover from our minds?

Making Memories without Learning— the Marilyn Monroe Experiment

The most challenging threshold to be crossed in the science of memory is to create the memory of an event that never occurred. This has been described as the Marilyn Monroe experiment—so named because, if given the chance, many in the community of older male scientists might be inclined to create a memory for themselves of a night spent in the company of the twentieth century's premier sex siren. Despite the frivolous name, success in such an experiment would be a tremendous achievement—the clearest proof that we understand the biological basis of memory.

There are strong hypotheses as to how the brain stores information. Currently, these remain hypotheses. We know that many of the molecules involved in synaptic plasticity are also involved in memory, and we also know that changes in the efficacy of synaptic transmission often accompany the formation of new memories. These findings, however, amount to circumstantial evidence; they reveal a correlation between phenomena, but they do not demonstrate causal relationships. In order to convincingly demonstrate that there is a causal relationship between synaptic plasticity and the formation and storage of memory, scientists must prove that synaptic plasticity is *necessary* for learning to occur and memories to be stored. To do this, they must selectively disable synaptic plasticity without compromising any other aspect of our physiology, and they must then observe a loss of memory. In some ways, the memory-erasure experiments already described may accomplish this; however, even if we can demonstrate necessity, there is still the issue of whether synaptic plasticity is *sufficient*: that is, the only process involved in learning and in storing memory. To test this hypothesis, we must show that changes in the efficacy of chemical synapses are sufficient to create a memory. This is where the Marilyn Monroe experiment comes in.

Demonstrating both necessity and sufficiency defines the process of scientific proof. Unfortunately, the experiments that test these two criteria are also the hardest to perform. In the case of synaptic plasticity and memory, they have not yet been done, and until such experiments have been conducted, the synaptic-plasticity hypothesis will remain a hypothesis. The difficulty in doing such an experiment is twofold: First, among the quadrillion (10^{15}) synapses in the human brain, scientists must identify the exact subset relevant to the formation and storage of a particular memory—no greater, no fewer. Second, they must be able to artificially alter those synapses in a manner that will mimic natural physiology. Currently, such experiments are being conducted in animal subjects—rats and mice, which have considerably fewer synapses than a human and in which invasive approaches can be used to manipulate synaptic strength.

Memory is studied in rodents by training in carefully designed arenas. The animals learn to find hidden platforms below the surface of opaque water, using spatial cues surrounding a pool, or they form fear memories of associations between particular places or sounds and an impending electrical shock. Other sorts of mazes and training apparatus test a multitude of rodent memories, each dependent on a different brain circuit and each used by scientists as models of particular types of human memories. We are beginning to understand which regions of the brain mediate each type of memory—episodic memories (what happened to us where and when), emotional memories (of frightening or rewarding events), procedural memories (of acquired skills). New technologies of optical imaging and microscopy allow us to observe real-time changes at synapses that occur as learning is under way. Once we can observe molecules interacting in the learning brain and measure the relevant patterns of neural activity in sufficient detail, we may be able to finally test the causal role of synaptic plasticity in memory.

In the near future, it seems that our best approximation of the Marilyn Monroe experiment will be to observe the exact pattern of change that occurs in the brain of a rat as it learns to associate a place with an electric shock: we may then attempt to erase the fear memory that results, which can be measured by observing the amount of time that the animal subsequently spends immobile, frozen in the place where it previously received the shock. Complete erasure of the memory, through propranolol application, should lead to the rat spending absolutely no time in this immobile state and instead freely exploring the place as if this were the first time it had ever been there. We would then be in a position to reinstall the memory by artificially recapitulating the exact changes we observed the first time the animal received the shock. The return of the freezing behavior would indicate the successful reinstallation of the rat's fear of impending shock.

This sequence of experimental proofs clearly is not entirely equivalent to the true Marilyn Monroe experiment, because the animal has actually experienced the key events initially. Nonetheless, it is conceptually similar and would accomplish the neuroscientist's aim of demonstrating the sufficiency of particular biological processes—in this case, synaptic plasticity—for the formation of memory. It is hoped that a greater understanding of the commonalities that exist from animal to animal when memories with a particular content are formed may eventually allow us to perform a true mimicry experiment, in which a memory is made de novo in a completely naive rat.

Designer Memory and the Abuse of Scientific Discovery

Whether we will actually shift such mimicry experiments into the realm of human beings and attempt the Marilyn Monroe experiment is another question entirely; first, we will have to ask ourselves if it is a worthwhile endeavor.

History suggests that society will eventually use available knowledge and technology for all possible ends, no matter how unsavory. A litany of scientific discoveries have first furthered our understanding of the universe but then have swiftly been put to darker uses. Witness the discovery of nuclear fission and the development of the atomic bomb, or Francis Galton's theory of eugenics and Josef Mengele's human experimentation in World War II. Such a progression is also apparent in the realm of lifestyle enhancement, as exemplified by cosmetic surgery, Viagra, and anabolic steroid abuse. Plastic surgery was originally developed with the benevolent intent of treating victims of serious burns. Viagra was developed as an angina treatment. Anabolic steroids were first used to rebuild muscle after injury and treat growth disorders. Inevitably, all three have been put to other uses for the sake of financial gain. One could argue that these lifestyle enhancements are not all to be deplored; nonetheless, the point is that society has a way of exploring all potential uses of a technology.

Nootropics, first developed for the treatment of Alzheimer's and other forms of dementia, are almost certain to find a wider use, enhancing the memory of the young and unafflicted. Memory-erasure technology may become a reality for the general populace. The creation of artificial memory is still the stuff of science fiction, but someday it may arrive. Then the ugly prospect of designer memory arises. We may become smarter, through chronic use of nootropics, and we may also be able to create a sanitized internal life through selective memory-reconstruction technology. Are we destined to lead a cosmeticized conscious existence, bereft of negative experiences and overflowing with positive false memories? Will those in power gain the extra weapon of mind control, the ability to wipe our memories of political misdeeds clean and implant edited ones in their place?

This dystopian vision presents scientists with an ethical dilemma: should we search for nootropics in order to treat cognitive disorders such as Alzheimer's disease, try to selectively erase the memories that give rise to PTSD, pursue the Marilyn Monroe experiment to finally test our hypotheses on the biological basis of memory—when we know that we run the risk of radically altering the texture of society in years to come? Some may argue that it is not the role of scientists to make ethical judgments about the potential impact of their work—that such decisions are the job of government, or the electorate, who should decide which scientific research is funded by public money and which is not. It is also argued that the complex ramifications of introducing any new knowledge or technology into the public domain are unpredictable enough that we cannot shy away from immediate gain for fear of future damage. Only time will tell whether nootropics and memory-reconstruction technologies are beneficial to society, but past experience suggests that we will come to use and abuse them as we have all other scientific discoveries. We have already started on a journey to a "smart" new world of nootropics, eternal sunshine, and Marilyn Monroe, whether we like it or not.

DEENA SKOLNICK WEISBERG

is a doctoral candidate in the Department of Psychology at Yale University, working toward her Ph.D. in developmental psychology. She received a B.S. in cognitive science from Stanford University in 2003, and she has also earned an M.S. and an M.Phil. from Yale. Weisberg's research focuses primarily on the cognitive skills underlying the creation and representation of nonreal scenarios—particularly stories, games of pretending, and counterfactual situations—and on how those skills mature in child development. She has also conducted research on how children learn new adjectives and on the ways in which adults misinterpret scientific findings, especially those of neuroscience. Her work has been published in a variety of scholarly journals, including Cognition, *the* Journal of Cognitive Neuroscience, *and* Science.

THE VITAL IMPORTANCE OF IMAGINATION

DEENA SKOLNICK WEISBERG

An extraordinary fact about our cognitive abilities is that we are not stuck in reality. We can travel back into the past (our own through memory, or the world's through history), forward into the future, and outward into the realm of fantasy, imagining possibilities we know are not real and may never be.

Even very young children are able to make this leap beyond reality into fantasy; the earliest examples of this ability are their games of "let's pretend." A child can have a tea party at which there is no tea, or move a pencil along a table while making car noises, knowing full well that the pencil is not a car. Indeed, researchers have found that children as young as two, novice pretenders, understand that pouring make-believe water on a stuffed bear drenches the bear, even though no water has actually been poured and the bear is not actually wet.[1]

[1] A. M. Leslie, "Pretending and Believing: Issues in the Theory of TOMM," *Cognition* 50(1994): 211–38.

As children get older, their understanding of the differences between reality and fantasy deepens, and they begin to interact in a mature way both with their pretend games and with stories not of their own creation. By at least age four, and often earlier, children have a firm grasp of the differences between real life and fantasy. They report that such creatures as witches and fairies don't really exist, and they know that a picture of a bear cooking in a kitchen doesn't depict a real event, although bears and kitchens are real.[2] They also understand the nature of the difference between reality and fantasy, knowing that they can eat a real cookie but not an imaginary one.[3]

Children would still understand all those things if they knew about only two categories of objects: those that are real and those that are fictional. If so, they would correctly understand that all real objects and people belong in the real world, but they would also believe that all fictional objects and people belong in one and the same fictional world.

This is not what adults believe. Adults create separate fictional worlds for different stories and generally do not expect characters from different stories to interact. Crossover stories—for example, *Shrek* or *The New Batman-Superman Adventures*—are clever and interesting to us precisely because they're unexpected; they violate the implicit rule that different stories

[2] For example, see P. Morison and H. Gardener, "Dragons and Dinosaurs: The Child's Capacity to Differentiate Fantasy from Reality," *Child Development* 49(1978): 642–48, and A. Samuels and M. Taylor, "Children's Ability to Distinguish Fantasy Events from Real-Life Events," *British Journal of Developmental Psychology* 12(1994): 417–27.
[3] H. M. Wellman and D. Estes, "Early Understanding of Mental Entities: A Reexamination of Childhood Realism," *Child Development* 57(1986): 910–23.

are isolated from each other. Under normal circumstances, characters stay in their own proper fictional worlds.

I designed a set of studies to find out whether children understand these divisions within the fantasy realm as adults do or whether they understand only the distinction between reality and fantasy. I asked four- and five-year-olds three types of questions about characters familiar to them, like Batman and SpongeBob SquarePants. First, I asked whether they understood that these characters are actually fictional. ("What do you think about Batman? Is he real or make-believe?") Second, I asked whether they understood that fictional characters in the same story are real to each other. ("What does Batman think about Robin? Does Batman think Robin is real or make-believe?") Finally, to test whether children organize stories into separate worlds, I asked whether characters from different stories could come into contact with each other. ("What does Batman think about SpongeBob? Does Batman think SpongeBob is real or make-believe?")

Children's responses on all three sets of questions were no different from the responses of adults asked the same questions. Like adults, children believe that Batman is fictional, that Batman and Robin are real to each other, and that Batman and SpongeBob are fictional to each other.[4] The remarkable thing about these responses is that children (and adults, too) probably never think explicitly about the relationships among various characters, but when I asked the children to categorize

[4] D. Skolnick and P. Bloom, "What Does Batman Think About SpongeBob? Children's Understanding of the Fantasy/Fantasy Distinction," *Cognition* 101(2006): B9–18.

such relationships, they had no trouble doing so. Children and adults alike impose an organizational structure on the stories they know, even if they may not realize they do.

This result prompted me to wonder whether children would similarly separate their pretend games. Pretend games are like stories, in that both take place outside reality and involve characters and sequences of actions—but they are unlike stories in that children have had more experience with them and exert direct control over them. Which would determine whether children separate the worlds of their pretend games—the similarities or the differences?

To answer this question, a research assistant and I created two pretend games, setting up a situation analogous to the two stories I'd asked about in the previous study. We used a set of colored blocks as pretend objects in these two games. For example, the first game involved a teddy bear who needed to take a bath. The child might pretend that one of the blocks was a bar of soap and scrub the bear with it. The second game involved a doll who needed to take a nap, and the child might give the doll a second block to be a pillow for her nap.

The crucial question was whether children think about their pretend games as separate worlds, the way they think about two stories, or whether there can be crossovers between the games. In this case, I wanted to test whether the child would be willing to take an object from one game and move it to the other. So I pretended that the bear, who had taken its bath, now needed to go to sleep. I said that the bear needed a pillow to go to sleep and asked the child to get a pillow for the bear. There were two responses a child might make to this request: either take the pillow from the doll's game or choose a new block to be a pillow for the bear. The former choice would indicate no separation between the two pretend games—that is, objects in one could be moved to another with no trouble. The latter choice would indicate a separation, by revealing the child's unwillingness to move an appropriate object from one game to another. The children in this study overwhelmingly preferred to use a new block rather than transfer objects between games, and they responded this way whether the two games were played in sequence or in parallel.

Thus it would seem that children treat their pretend games in the same way as the stories they hear; these two types of fictional scenarios share the same organizational structure. The existence of a common structure suggests that the ability to think about stories and pretend games is supported by the same cognitive mechanism, which I call the *what-if* mechanism (WIM). This mechanism, as its name suggests, is a cognitive apparatus that allows us to ask "What if . . . ?" to explore possibilities that do not currently exist in reality. We do this in our imaginations, setting up scenarios and letting them play out, and we also do it with the help of authors and filmmakers, as we follow their explorations of the possibilities they have set up. What if I could turn invisible? What if there existed a ring of power, and what if that ring were to fall into the hands of a hobbit? What if a proud young woman named Scarlett O'Hara had lived in the American South at the time of the Civil War? In these cases and many others, all we need to do in order to transport ourselves outside current sensory reality is to ask "What if?" engaging the WIM.

I believe that all our various forays outside reality are supported by this mechanism, whether we are listening to stories, playing pretend games, daydreaming, imagining the future, or trying to work out what could have happened in the past. In all these cases, we need to create a representation of something outside reality. Furthermore, we know we're moving outside reality and we can report that these representations aren't real. In reality, there is a pencil; in the pretend game, there's a car. In reality, there is no such person as Bruce Wayne; in the story, Bruce Wayne exists and leads a secret life as a crime-fighting superhero. Exactly the same sort of reasoning is involved when we think about an alternative past (counterfactual reasoning) or when we plan for the future (hypothetical reasoning). In reality, events unfolded one way; in my imagination, I work through what might have happened if they had unfolded another way. In reality, I can see only the present; in my imagination, I can set in motion possible future scenarios to help me predict what will happen and what my best course of action is. I call all these types of representations *fictional worlds*, to capture their two defining properties: they're not real, and we know they're not real.

The main goal of my research is to discover the nature of the what-if mechanism and how it allows us to create and comprehend fictional worlds. The two studies I conducted have revealed at least one of its signatures: multiple fictional worlds are kept separate from one another. Those studies found this multiple-world structure both for stories and for pretend games, and I am hopeful that my future studies will find it for counterfactual and hypothetical scenarios.

Besides separating fictional worlds from each other, another job of the WIM is to navigate the boundary between reality and fiction. In some ways, this boundary is rigid. Characters like

Batman are fictional and shouldn't migrate into reality; a block that one pretends is a cookie shouldn't be mistaken for a real cookie. But in other ways the boundary is more fluid. For one thing, we take many aspects of reality with us when we enter a fictional world. Batman might exist as a crime-fighting super-hero, or time travel might be possible, but two plus two still equals four, water still turns to ice when it freezes, and it is still impossible for something to be colored blue and not colored blue at the same time. The laws of mathematics, science, and logic still hold, even in what appears to be a radically different, fictional world. The WIM must thus be able to project an appro-priate set of real-world facts and rules into a fictional realm.

Moreover, we take many aspects of a fictional world with us when we move out of fiction and back into reality. This might seem like a strange claim, but consider that we learn quite a lot about reality by reading books and watching movies and TV shows—including, importantly, things we would never other-wise learn. When we read or see *Othello*, we learn about the destructive power of jealousy, even if we've never been driven mad by it and likely never will be. When we watch an episode of *CSI*, we learn what collecting and analyzing DNA evidence is like, even if we've never seen or used those techniques in real life. And if someone told you he was going to find out what life was like in nineteenth-century Russia by reading a Tolstoy novel, you wouldn't bat an eyelash. Learning about reality from fiction happens all the time, whether it's understanding emo-tional truths, getting a feel for a country or an era, or picking up some practical information.

This is precisely the property of fictional-world cognition that makes it so useful to us—and so useful to children as they develop. There's a lot of reality we can't explore directly, especially as children. But by using our imaginations, we can still learn about things we don't actually experience. In fact, some researchers speculate that this is what pretending is all about: giving children an opportunity to play other social roles or explore other aspects of their personalities in a safe, separate space. Children may not learn practical facts from a pretend game, as they can from a story, but they probably do learn social and emotional facts—like how to interact in certain scenarios, or how it feels to do certain things without experiencing the scenario for real.

Every day there are problems we need to solve, and we use the WIM to help us. Maybe the car won't start, and we need to find out why. We can run through fictional scenarios: What if the battery is dead? What if the transmission is broken? For each possibility we come up with, we use the WIM to set a scenario in motion to see what would follow if that possibility were true. We can then move back into reality to test those possibilities, using the information from our imagined scenarios to help us plan. If the battery is dead, I can get my car jump-started, but if the transmission is broken, jump-starting won't help. Even these simple calculations require imagination, require deploying the WIM, require moving outside reality. Reality offers no help; the car won't start, and that's all the information I have. I can get more information from reality alone by exhaustively performing a wide variety of actions to try to get the car started, but that's not a practical way to collect data: by temporarily sidestepping reality and creating an imagined scenario in which I know what's wrong, I make the best use of my time and resources to solve the problem facing me.

This function of the WIM is even more essential for children, who have so much more of reality to puzzle through. It would be impossible for them to perform every possible action they could take at any point. Like adults, they need to move outside reality and work through what might happen if they were to do a particular thing. If the result is favorable, they can go ahead and try it; if not, they can think of a different possible action and they won't have wasted time on something that isn't going to work.

Basically, the WIM is an engine for learning. Although its primary function is to get us outside reality, these fictional ventures provide us with a new window onto reality. It might be fun to play a pretend game or listen to a story, but in the background the WIM is doing its work of pulling useful information from these fictional worlds into reality, expanding our practical, emotional, and social knowledge. The trick is to make sure it's pulling the right information. This is a major question for my future research: how do we know what's appropriate to take from fiction into reality? Is this something we know automatically, the way we automatically know how to create fictional worlds? Or is it something that requires experience and learning?

I suspect the answer lies somewhere in between. Children probably know implicitly that they can get useful real world information from a fictional scenario. This is what lets them perform productive actions in the world based on their ability to plan for the future, and it's what lets them work through real-world issues in their pretend games. But there are also nuances with respect to what adults take from a fictional world. For instance, Harry Potter isn't real, and we shouldn't bring the character himself from fiction into reality. Even children know this much. But what about the house where he lives in England, no. 4 Privet Drive? Could this street exist in reality? What about his mentor, Professor Dumbledore? He doesn't really exist, but could there be someone similar to him in reality? Perhaps, but similar to him in what ways? A real person wouldn't have Dumbledore's magical abilities, but which of his other physical or psychological traits are potentially real?

Adults have subtle intuitions about the answers to those questions that children likely don't share. Children's intuitions are probably more extreme than adults'; that is, they may take too much from fiction into reality—or too little. Children may see the boundaries between reality and fiction as more fluid than adults do; they may pull nearly everything from fiction into reality and hence believe in the existence of strange entities or in the operation of strange rules that don't actually exist or operate. Conversely, they may see fiction as *more* separate from reality than adults do and not understand that even in stories, most of what is depicted about human interactions or physical phenomena could easily be real.

I hope to shed light on these issues. But what's clear for now is just how important the WIM is to children. This basic imaginative skill, which lets us pop temporarily outside reality to work through various possibilities, is one of the primary tools children use to make sense of the world.

DAVID M. EAGLEMAN

earned his undergraduate degree in British and American lit-erature at Rice University and Oxford University and a Ph.D. in neuroscience at Baylor College of Medicine in 1998. He is now the director of BCM's Laboratory for Perception and Action, whose long-range goal is to understand the neural mechanisms of time perception. Eagleman also directs BCM's Initiative on Law, Brains, and Behavior, which seeks to deter-mine how new discoveries in neuroscience will change our laws and criminal justice system. Eagleman is the author of Sum: Forty Tales from the Afterlives *as well as the upcoming* Plasticity: How the Brain Reconfigures Itself.

BRAIN TIME

DAVID M. EAGLEMAN

At some point, the Mongol military leader Kublai Khan (1215–94) realized that his empire had grown so vast that he would never be able to see what it contained. To remedy this, he commissioned emissaries to travel to the empire's distant reaches and convey back news of what he owned. Since his messengers returned with information from different distances and traveled at different rates (depending on weather, conflicts, and their fitness), the messages arrived at different times. Although no historians have addressed this issue, I imagine that the Great Khan was constantly forced to solve the same problem a human brain has to solve: what events in the empire occurred in which order?

Your brain, after all, is encased in darkness and silence in the vault of the skull. Its only contact with the outside world is via the electrical signals exiting and entering along the super-highways of nerve bundles. Because different types of sensory information (hearing, seeing, touch, and so on) are processed at different speeds by different neural architectures, your brain faces an enormous challenge: what is the best story that can be constructed about the outside world?

The days of thinking of time as a river—evenly flowing, always advancing—are over. Time perception, just like vision, is a construction of the brain and is shockingly easy to manipulate experimentally. We all know about optical illusions, in which things appear different from how they really are; less well known is the world of temporal illusions. When you begin to look for temporal illusions, they appear everywhere. In the movie theater, you perceive a series of static images as a smoothly flowing scene. Or perhaps you've noticed when glancing at a clock that the second hand sometimes appears to take longer than normal to move to its next position—as though the clock were momentarily frozen.

More subtle illusions can be teased out in the laboratory. Perceived durations are distorted during rapid eye movements, after watching a flickering light, or simply when an "oddball" is seen in a stream of repeated images. If we inject a slight delay between your motor acts and their sensory feedback, we can later make the temporal order of your actions and sensations appear to reverse. Simultaneity judgments can be shifted by repeated exposure to nonsimultaneous stimuli. And in the laboratory of the natural world, distortions in timing are induced by narcotics such as cocaine and marijuana or by such disorders as Parkinson's disease, Alzheimer's disease, and schizophrenia.

Try this exercise: Put this book down and go look in a mirror. Now move your eyes back and forth, so that you're looking at your left eye, then at your right eye, then at your left eye again. When your eyes shift from one position to the other, they take time to move and land on the other location. But here's the kicker: you never see your eyes move. What is happening to the time gaps during which your eyes are moving? Why do you feel as though there is no break in time while you're changing your eye position? (Remember that it's easy to detect someone else's eyes moving, so the answer cannot be that eye movements are too fast to see.)

All these illusions and distortions are consequences of the way your brain builds a representation of time. When we examine the problem closely, we find that "time" is not the unitary phenomenon we may have supposed it to be. This can be illustrated with some simple experiments: for example, when a stream of images is shown over and over in succession, an oddball image thrown into the series appears to last for a longer period, although presented for the same physical duration. In the neuroscientific literature, this effect was originally termed a subjective "expansion of time," but that description begs an important question of time representation: when durations dilate or contract, does time *in general* slow down or speed up during that moment? If a friend, say, spoke to you during the oddball presentation, would her voice seem lower in pitch, like a slowed-down record?

If our perception works like a movie camera, then when one aspect of a scene slows down, everything should slow down. In the movies, if a police car launching off a ramp is filmed in slow motion, not only will it stay in the air longer but its siren will blare at a lower pitch and its lights will flash at a lower frequency. An alternative hypothesis suggests that different temporal judgments are generated by different neural mechanisms—and while they often agree, they are not required to. The police car may seem suspended longer, while the frequencies of its siren and its flashing lights remain unchanged.

Available data support the second hypothesis.[1] Duration distortions are not the same as a unified time slowing down, as it does in movies. Like vision, time perception is underpinned by a collaboration of separate neural mechanisms that usually work in concert but can be teased apart under the right circumstances.

This is what we find in the lab, but might something different happen during real-life events, as in the common anecdotal report that time "slows down" during brief, dangerous events such as car accidents and robberies? My graduate student Chess Stetson and I decided to turn this claim into a real scientific question, reasoning that if time as a single unified entity slows down during fear, then this slow motion should confer a higher temporal resolution—just as watching a hummingbird in slow-motion video allows finer temporal discrimination upon replay at normal speed, because more snapshots are taken of the rapidly beating wings.

We designed an experiment in which participants could see a particular image only if they were experiencing such enhanced temporal resolution. We leveraged the fact that the visual brain integrates stimuli over a small window of time: if two or more images arrive within a single window of integration (usually under one hundred milliseconds), they are perceived as a single image. For example, the toy known as a thaumatrope

[1] V. Pariyadath and D. M. Eagleman, "The Effect of Predictability on Subjective Duration," *PLoS ONE* (2007).

may have a picture of a bird on one side of its disc and a picture of a tree branch on the other; when the toy is wound up and spins so that both sides of the disc are seen in rapid alternation, the bird appears to be resting on the branch. We decided to use stimuli that rapidly alternated between images and their negatives. Participants had no trouble identifying the image when the rate of alternation was slow, but at faster rates the images perceptually overlapped, just like the bird and the branch, with the result that they fused into an unidentifiable background.

To accomplish this, we engineered a device (the perceptual chronometer) that alternated randomized digital numbers and their negative images at adjustable rates. Using this, we measured participants' threshold frequencies under normal, relaxed circumstances. Next, we harnessed participants to a platform that was then winched fifteen stories above the ground. The perceptual chronometer, strapped to the participant's forearm like a wristwatch, displayed random numbers and their negative images alternating just a bit faster than the participant's determined threshold. Participants were released and experienced free fall for three seconds before landing (safely!) in a net. During the fall, they attempted to read the digits. If higher temporal resolution were experienced during the free fall, the alternation rate should appear slowed, allowing for the accurate reporting of numbers that would otherwise be unreadable.[2]

---

[2] A critical point is that the speed at which one can discriminate alternating patterns is not limited by the eyes themselves, since retinal ganglion cells have extremely high temporal resolution. For more details on this study, see C. Stetson et al., "Does Time Really Slow Down During a Frightening Event?" *PLoS ONE* (2007).

The result? Participants weren't able to read the numbers in free fall any better than in the laboratory. This was not because they closed their eyes or didn't pay attention (we monitored for that) but because they could not, after all, see time in slow motion (or in "bullet time," like Neo in *The Matrix*). Nonetheless, their perception of the elapsed duration itself was greatly affected. We asked them to retrospectively reproduce the duration of their fall using a stopwatch. ("Re-create your free-fall in your mind. Press the stopwatch when you are released, then press it again when you feel yourself hit the net.") Here, consistent with the anecdotal reports, their duration estimates of their own fall were a third greater, on average, than their re-creations of the fall of others.

How do we make sense of the fact that participants in free fall reported a duration expansion yet gained no increased discrimination capacities in the time domain during the fall? The answer is that time and memory are tightly linked. In a critical situation, a walnut-size area of the brain called the amygdala kicks into high gear, commandeering the resources of the rest of the brain and forcing everything to attend to the situation at hand. When the amygdala gets involved, memories are laid down by a secondary memory system, providing the later flash-bulb memories of post-traumatic stress disorder. So in a dire situation, your brain may lay down memories in a way that makes them "stick" better. Upon replay, the higher density of data would make the event appear to last longer. This may be why time seems to speed up as you age: you develop more compressed representations of events, and the memories to be read out are correspondingly impoverished. When you are a child, and everything is novel, the richness of the memory gives the impression of increased time passage—for example, when looking back at the end of a childhood summer.

To further appreciate how the brain builds its perception of time, we have to understand where signals are in the brain, and when. It has long been recognized that the nervous system faces the challenge of *feature-binding*—that is, keeping an object's features perceptually united, so that, say, the redness and the squareness do not bleed off a moving red square. That feature-binding is usually performed correctly would not come as a surprise were it not for our modern picture of the mammalian brain, in which different kinds of information are processed in different neural streams. Binding requires coordination—not only among different senses (vision, hearing, touch, and so on) but also among different features within a sensory modality (within vision, for example: color, motion, edges, angles, and so on).

But there is a deeper challenge the brain must tackle, without which feature-binding would rarely be possible. This is the problem of *temporal*-binding: the assignment of the correct timing of events in the world. The challenge is that different stimulus features move through different processing streams and *are processed at different speeds*. The brain must account for speed disparities between and within its various sensory channels if it is to determine the timing relationships of features in the world.

What is mysterious about the wide temporal spread of neural signals is the fact that humans have quite good resolution when making temporal judgments. Two visual stimuli can be accurately deemed simultaneous down to five milliseconds, and their order can be assessed down to twenty-millisecond resolutions. How is the resolution so precise, given that the signals are so smeared out in space and time?

To answer this question, we have to look at the tasks and resources of the visual system. As one of its tasks, the visual system—couched in blackness, at the back of the skull—has to get the timing of outside events correct. But it has to deal with the peculiarities of the equipment that supplies it: the eyes and parts of the thalamus. These structures feeding into the visual cortex have their own evolutionary histories and idiosyncratic circuitry. As a consequence, signals become spread out in time from the first stages of the visual system (for example, based on how bright or dim the object is).

So if the visual brain wants to get events correct timewise, it may have only one choice: *wait for the slowest information to arrive.* To accomplish this, it must wait about a tenth of a second. In the early days of television broadcasting, engineers worried about the problem of keeping audio and video signals synchronized. Then they accidentally discovered that they had around a hundred milliseconds of slop: as long as the signals arrived within this window, viewers' brains would automatically resynchronize the signals; outside that tenth-of-a-second window, it suddenly looked like a badly dubbed movie.

This brief waiting period allows the visual system to discount the various delays imposed by the early stages; however, it has the disadvantage of pushing perception into the past. There is a distinct survival advantage to operating as close to the present as possible; an animal does not want to live too far in the past. Therefore, the tenth-of-a-second window may be the smallest delay that allows higher areas of the brain to account for the delays created in the first stages of the system while still operating near the border of the present. This window of delay means that awareness is postdictive, incorporating data from a window of time after an event and delivering a retrospective interpretation of what happened.[3]

Among other things, this strategy of waiting for the slowest information has the great advantage of allowing object recognition to be independent of lighting conditions. Imagine a striped tiger coming toward you under the forest canopy, passing through successive patches of sunlight. Imagine how difficult recognition would be if the bright and dim parts of the tiger caused incoming signals to be perceived at different times. You would perceive the tiger breaking into different space-time fragments just before you became aware that you were the tiger's lunch. Somehow the visual system has evolved to reconcile different speeds of incoming information; after all, it is advantageous to recognize tigers regardless of the lighting.

This hypothesis—that the system waits to collect information over the window of time during which it streams in—applies not only to vision but more generally to all the other senses. Whereas we have measured a tenth-of-a-second window of postdiction in vision, the breadth of this window may be different for hearing or touch. If I touch your toe and your nose at the same time, you will feel those touches as simultaneous. This is surprising, because the signal from your nose reaches your brain well before the signal from your toe. Why didn't you

---

[3] We introduced the term *postdiction* in 2000 to describe the brain's act of collecting information well after an event and then settling on a perception (D. M. Eagleman and T. J. Sejnowski, "Motion Integration and Postdiction in Visual Awareness," *Science* 287(2000): 2036–38).

feel the nose-touch when it first arrived? Did your brain wait to see what else might be coming up the pipeline of the spinal cord until it was sure it had waited long enough for the slower signal from the toe? Strange as that sounds, it may be correct.

It may be that a unified polysensory perception of the world has to wait for the slowest overall information. Given conduction times along limbs, this leads to the bizarre but testable suggestion that tall people may live further in the past than short people. The consequence of waiting for temporally spread signals is that perception becomes something like the airing of a live television show. Such shows are not truly live but are delayed by a small window of time, in case editing becomes necessary.

Waiting to collect all the information solves part of the temporal-binding problem, but not all of it. A second problem is this: if the brain collects information from different senses in different areas and at different speeds, how does it determine how the signals are supposed to line up with one another? To illustrate the problem, snap your fingers in front of your face. The sight of your fingers and the sound of the snap appear simultaneous. But it turns out that impression is laboriously constructed by your brain. After all, your hearing and your vision process information at different speeds. A gun is used to start sprinters, instead of a flash, because you can react faster to a bang than to a flash. This behavioral fact has been known since the 1880s and in recent decades has been corroborated by physiology: the cells in your auditory cortex can change their firing rate more quickly in response to a bang than your visual cortex cells can in response to a flash.

The story seems as though it should be wrapped up here.

Yet when we go outside the realm of motor reactions and into the realm of perception (what you *report* you saw and heard), the plot thickens. When it comes to awareness, your brain goes through a good deal of trouble to perceptually synchronize incoming signals that were synchronized in the outside world. So a firing gun will seem to you to have banged and flashed at the same time. (At least when the gun is within thirty meters; past that, the different speeds of light and sound cause the signals to arrive too far apart to be synchronized.)

But given that the brain received the signals at different times, how can it know what was supposed to be simultaneous in the outside world? How does it know that a bang didn't really happen before a flash? It has been shown that the brain constantly recalibrates its expectations about arrival times. And it does so by starting with a single, simple assumption: if it sends out a motor act (such as a clap of the hands), all the feedback should be assumed to be simultaneous and any delays should be adjusted until simultaneity is perceived. In other words, the best way to predict the expected relative timing of incoming signals is to interact with the world: each time you kick or touch or knock on something, your brain makes the assumption that the sound, sight, and touch are simultaneous.

While this is a normally adaptive mechanism, we have discovered a strange consequence of it: Imagine that every time you press a key, you cause a brief flash of light. Now imagine we sneakily inject a tiny delay (say, two hundred milliseconds) between your key-press and the subsequent flash. You may not even be aware of the small, extra delay. However, if we suddenly remove the delay, you will now believe that the flash occurred *before* your key-press, an illusory reversal of action and sensation.

Your brain tells you this, of course, because it has adjusted to the timing of the delay.

Note that the recalibration of subjective timing is not a party trick of the brain; it is critical to solving the problem of causality. At bottom, causality requires a temporal order judgment: did my motor act come before or after that sensory signal? The only way this problem can be accurately solved in a multisensory brain is by keeping the expected time of signals well calibrated, so that "before" and "after" can be accurately determined even in the face of different sensory pathways of different speeds.

It must be emphasized that everything I've been discussing is in regard to conscious awareness. It seems clear from preconscious reactions that the motor system does not wait for all the information to arrive before making its decisions but instead acts as quickly as possible, *before* the participation of awareness, by way of fast subcortical routes. This raises a question: what is the use of perception, especially since it lags behind reality, is retrospectively attributed, and is generally outstripped by automatic (unconscious) systems? The most likely answer is that perceptions are representations of information that cognitive systems can work with later. Thus it is important for the brain to take sufficient time to settle on its best interpretation of what just happened rather than stick with its initial, rapid interpretation. Its carefully refined picture of what just happened is all it will have to work with later, so it had better invest the time.

Neurologists can diagnose the variety of ways in which brains can be damaged, shattering the fragile mirror of perception into unexpected fragments. But one question has gone mostly unasked in modern neuroscience: what do disorders of *time* look like? We can roughly imagine what it is like to lose color vision, or hearing, or the ability to name things. But what would it feel like to sustain damage to your time-construction systems?

Recently, a few neuroscientists have begun to consider certain disorders—for example, in language production or reading—as potential problems of timing rather than disorders of language as such. For example, stroke patients with language disorders are worse at distinguishing different durations, and reading difficulties in dyslexia may be problems with getting the timing right between the auditory and visual representations.[4] We have recently discovered that a deficit in temporal-order judgments may underlie some of the hallmark symptoms of schizophrenia, such as misattributions of credit ("My hand moved, but I didn't move it") and auditory hallucinations, which may be an order reversal of the generation and hearing of normal internal monolog.

As the study of time in the brain moves forward, it will likely uncover many contact points with clinical neurology. At present, most imaginable disorders of time would be lumped into a classification of dementia or disorientation, catchall diagnoses that miss the important clinical details we hope to discern in coming years.

Finally, the more distant future of time research may change our views of other fields, such as physics. Most of our

[4] R. Efron, "Temporal Perception, Aphasia, and Deja Vu," *Brain* 86(1963): 403–24; M. M. Merzenich et al., "Temporal Processing Deficits of Language-Learning Impaired Children Ameliorated by Training," *Science* 271, no. 5245 (1996): 77–81.

current theoretical frameworks include the variable t in a New-tonian, river-flowing sense. But as we begin to understand time as a construction of the brain, as subject to illusion as the sense of color is, we may eventually be able to remove our perceptual biases from the equation. Our physical theories are mostly built on top of our filters for perceiving the world, and time may be the most stubborn filter of all to budge out of the way.

VANESSA WOODS,

author of It's Every Monkey for Themselves, *is an award-winning journalist. She has a double degree in biology and English from the University of New South Wales and received a master's degree in science communication from the Centre for the Public Awareness of Science at the Australian National University in 2004. She is a researcher with the Hominoid Psychology Research Group and studies the psychology of bonobos and chimpanzees in Africa.*

BRIAN HARE

is an anthropologist and an assistant professor in the Department of Biological Anthropology and Anatomy at Duke University. He studied anthropology, psychology, and human and natural ecology at Emory University and received his Ph.D. from Harvard University in 2004. His research centers on human cognitive evolution, and his experience in the field includes working on a fox farm in Siberia, chasing chimpanzees through the jungle of Uganda, and helping orphaned bonobos in the Democratic Republic of the Congo.

OUT OF OUR MINDS

How Did *Homo sapiens* Come Down from the Trees, and Why Did No One Follow?

VANESSA WOODS AND BRIAN HARE

Mikeno sits with his chin resting on his right hand, in a startling imitation of Rodin's *Thinker*. His left arm is thrown over his knee, and his eyes are slightly out of focus, as though he's deep in thought. With his black hair parted carefully down the middle and his rosy pink lips, Mikeno looks human. But he isn't. Mikeno is a bonobo—an inhabitant of Lola Ya Bonobo, one of a number of African sanctuaries for apes orphaned by the bush-meat trade, this one in the Democratic Republic of the Congo.

Bonobos share more DNA (98.7 percent) with us than they do with gorillas—enough so that under his glossy black hair Mikeno has the body of a young athlete, complete with chiseled biceps and a developing six-pack. The question is: where among the three billion nucleotides of his genome is the 1.3 percent that makes Mikeno a bonobo instead of a human?

We have been seeking to define our humanity for thousands of years. Plato described a human being as a featherless creature that walks on two legs; in response, Diogenes turned up at one of Plato's lectures holding a plucked chicken. Other definitions have come and gone: Only humans use tools. Only humans intentionally murder one another. Only humans have souls. Like mirages in the desert, the definitions are always shifting.

In the six million years since hominids split from the evolutionary ancestor we share with chimpanzees and bonobos, something happened to our brains that allowed us to become master cooperators, accumulate knowledge at a rapid rate, and manipulate tools to colonize almost every corner of the planet. In evolutionary time, our progress has been swift and ruthless. What allowed us to come down from the trees, and why?

Are You Thinking What I'm Thinking?

When children turn four, they start to wonder what other people are thinking. For instance, if you show a four-year-old a packet of gum and ask what's inside, she'll say, "Gum." You open the packet and show her that inside there's a pencil instead of gum. If you ask her what her mother, who's waiting outside, will think is in the packet once it's been reclosed, she'll say, "Gum," because she knows her mother hasn't seen the pencil. But children under the age of four will generally say their mother will think there's a pencil inside—because children this young cannot yet escape the pull of the real world. They think everyone knows what they know, because they cannot model someone else's mind and in this case realize that someone must see something in order to know it.

This ability to think about what others are thinking about is called having a theory of mind.

Humans constantly want to know what others are thinking: *Did he see me glance at him? Does that beautiful woman want to approach me? Does my boss know I was not at my desk?* A theory of mind allows for complex social behaviors, such as military strategies, and the formation of institutions, such as governments.

Throughout the 1990s, scientists ran dozens of pioneering experiments in an attempt to determine whether chimpanzees—who, like bonobos, share 98.7 percent of our DNA—possess a theory of mind. An experiment conducted by Daniel Povinelli of the University of Louisiana at Lafayette gave chimpanzees the choice of using a visual gesture to request food from someone who was blindfolded, someone with a bucket over his head, someone whose hands were over his eyes, or someone who could actually see them. The chimps didn't discriminate; they made the begging gesture at people who obviously couldn't see them just as often as they begged from people who were looking straight at them. If chimpanzees have no theory of mind, which this set of findings suggested, then that could be what distinguishes humans from other animals.

That was before Brian and two colleagues, Josep Call and Michael Tomasello, began working with Jahaga, a female chimpanzee at the Wolfgang Köhler Primate Research Center of the Leipzig Zoo. The experiment went like this: In a room at the center, you sit behind a Perspex panel with a tray extending from it that holds a banana. Jahaga sees the banana. She can also see you watching and knows that if you see her coming, you'll pull in the tray, because you've already kept food from her like that. Instead of simply rushing for the banana, Jahaga casually walks to the back of the room, as though she didn't want your measly banana and was bored by the whole game. She continues along the back wall, slinking around a partition until she's out of sight. Then, when she knows the partition is blocking your view of her, she walks low and fast behind it and swipes the banana off the tray.

This was the first experiment to investigate whether chimpanzees will actively deceive another individual based on what that individual can or cannot see. Deception can be one important test of whether or not you possess a theory of mind, because, in many cases, in order to deceive someone you have to know what they're likely to be thinking and then try to manipulate the situation such that their thinking changes in your favor. Jahaga's behavior in this experiment—and later that of other chimpanzees—seemed deceptive, not just because she slinked to a place where she knew you couldn't see her (that is, she was sensitive to what you were thinking) but also because she seemed to be deceptive about being deceptive: she looked as though she were pretending not to be interested in the banana (that is, she may have been trying to manipulate what you thought about her intent).

After Jahaga, a whole range of experiments have shown that in a number of contexts chimpanzees do think about what others are thinking about. Low-ranking chimpanzees will always go for the food that's hidden from a dominant chimpanzee's view, because they know the dominant has not seen it. If you suddenly look up, a chimpanzee will follow your gaze, wondering what you've seen. If you delay giving chimpanzees food, either by teasing them or accidentally dropping it, they know when you're being intentionally mean, and they act more frustrated than they do when you're just being clumsy. But does this mean that chimpanzees have the same theory of mind that we do?

Point It Out

Even though Jahaga and other chimpanzees exhibit a sophisticated theory of mind on one level, on another level they're hopeless. If you hide a banana under one of two cups in such a way that Jahaga cannot see which cup you've chosen, and then you point to the cup where the banana is, Jahaga can't use your gesture to find it. You can tap on the cup, put a bright-colored block on it, maybe even dance around it, but Jahaga won't pick the correct cup any more often than she picks the wrong one. Dozens of trials later, she might start guessing the pattern, but if you change the cue from pointing to, say, tapping, she doesn't realize that the new cue will help her find the food. She has to learn to make use of your new gesture all over again.

However, human children under the age of two can use your pointing to find food. Even if you just *look* at the correct cup, children will follow your gaze and use it to gain information about what you know. They understand that you're trying to help them by communicating the location of the hidden goodie.

From these types of experiments with chimpanzees, it seems reasonable to conclude that using communicative gestures is something that evolved in our species after our lineage split from the other apes. Perhaps sharing information in this way enabled early humans to develop a much more complex form of culture than that seen in other animals. But if that's so, then how might such an ability have evolved in the first place?

Go Fetch

Oreo was the best dog any kid could wish for. He would take you to your friend's house and sit outside until it was time to ride your bike home again. He would let you give him as many hugs as you wanted when you were at an age when it wasn't cool to hug anyone except your dog. Most important, Oreo loved to play fetch. He would play fetch until your arm fell off, because he could easily carry three tennis balls in his mouth at once. The problem was, he usually couldn't keep up with where all the balls were going; after collecting the first two, he wouldn't have a clue where the third one had landed. After a few moments of frantic searching, he would race back to eye you, panting expectantly, waiting. If you pointed in the right direction, he would be back seconds later, with all three balls covered with slobber and ready for throwing again.

Anyone with a dog knows that when they want something and they know that you know where it is, they will watch your body language like a hawk for the slightest clue. Sure enough, when Brian and colleagues played the cup game with a myriad of dogs, they could point to, gaze at, or tap with a toe on the hiding place and the dogs would immediately find the hidden treat (and not because of their powerful noses—in these experiments, dogs cannot determine which cup hides the food without a visual cue).

Why does an animal like a dog succeed where our closest living relative fails?

One idea is that dogs live with us, so over thousands of hours of interacting with us, they learn to read our body language. Another idea is that the pack lifestyle and cooperative hunting of wolves, the canids from which all dogs evolved, made all canids, dogs included, more in tune with social cues.

To test the first idea, you need to play with puppies. If nine-week-old puppies pass the cup test, then perhaps reading human gestures isn't something dogs learn as they grow older but something they're born with. Brian and colleagues found that such puppies passed the test, but there was still the question of whether their first nine weeks had been enough to pick up human communicative gestures. So puppies reared in a kennel, with very little exposure to humans, were tested, too. The kennel puppies passed.

As for the second idea, you need to spend some time with the big bad wolves. When Brian and colleagues tested wolves at a wolf sanctuary and compared their accomplishments with those of a group of pet dogs, it became obvious that wolves were no better than chimpanzees at acting on human social cues. Thus it seems that dogs must have evolved to act on human social cues within the last forty thousand years—that is, since they split from their wolf ancestor through the process of being domesticated. The implications are exciting: a social skill that is an important developmental basis for human culture, cooperation, and language—a precursor and component of the human theory of mind—may have evolved in the dog as a result of interacting with us over many generations. Could it really be that domestication can lead to such a change in problem-solving abilities? So it would seem, but to test this idea you have to go to the middle of Siberia.

Clever Fox

The train ride from Moscow to Novosibirsk in summer is two days of green meadows filled with bright flowers. Once you get to Novosibirsk, you journey another half hour or so to Akademgorodok, the home of one of the greatest experiments in modern genetics.

Dmitri Belyaev was fired from a research laboratory in Moscow because his Mendelian view of genetics conflicted with that of Trofim Lysenko, the great Soviet scientist. Belyaev was lucky that his punishment ended with losing his Moscow job; under Stalin, dissent from Lysenko's theories of environmentally acquired inheritance was against the law, and many prominent scientists died in the Gulag. In 1958, Belyaev moved to Novosibirsk, where he became director of the Institute of Cytology and Genetics and, in the following year, began breeding 130 silver foxes in a kind of Mendelian experiment. He put one group under severe selection pressure using a simple method: those foxes that approached an experimenter lived to breed for another generation; those that snarled at humans or showed aggression toward them were turned into fur coats. The other group, a control, was bred randomly with regard to how they behaved toward humans.

After only forty generations, the selected foxes began to display changes you (and Darwin, too) might think would take millions of years to evolve. As expected, they became incredibly friendly toward humans. Whenever they saw people, they barked, wagged their tails, sniffed the people, and licked their faces. But even stranger were the physical changes, which occurred at a higher frequency than in the control group. The ears of the selected foxes became floppy. Their tails turned curly. Their coats lost their camouflage and became spotty, with a star pattern appearing on the forehead. Their skulls became smaller. In short, they looked and behaved remarkably like their close relative the domestic dog.

Now came the big test. If dogs had acquired social skills in the process of domestication, then perhaps the selected silver foxes acquired those skills, too.

And they did. Domesticated silver foxes could read human body language as well as any dog. The control lineage could not.

The skill of silver foxes at reading human social cues is a crucial piece of the puzzle. People (including the authors) had supposed that the unusual social skills found in dogs had probably evolved because smarter dogs had been more likely to survive and reproduce during domestication. But Belyaev's foxes weren't bred to be smarter than the average fox, just friendlier. It seems that the selected foxes are more skilled at reading human cues as a by-product of a loss of fear of humans, which was replaced by an intense interest in interacting with us. The social skills of dogs may have evolved through a similar process during their domestication. In order to avail themselves of garbage around human settlements, protodogs had to lose their fear of us. Subsequently, and by accident, while interacting with us, they began deploying the social skills they were using to interact with one another—as if we were just part of the pack.

Most important (and controversial), something similar may

have happened in human evolution. Instead of getting a jump start with the most intelligent hominids surviving to produce the next generation, as is often suggested, it may have been the more sociable hominids—because they were better at solving problems together—who achieved a higher level of fitness and allowed selection to favor more sophisticated problem solving over time. Humans got their smarts only because we got friendlier first.

The Chimpanzee Deficit

Cooperation is a cornerstone of human achievement, in part dependent on our sophisticated theory of mind and use of social cues. But humans are not the only species to be skilled cooperators. What is it about humans that makes us such flexible cooperators? Or, put another way: what goes wrong with chimpanzee cooperation? They live in highly social groups, hunt food together, maintain political relationships. What stops them from becoming as flexible as humans (or dogs, for that matter) at solving problems involving cooperation and communication?

Ngamba Island Chimpanzee Sanctuary is a sprawling hundred acres of primary forest in the middle of Lake Victoria, in Uganda. On a clear day, you can hear the pant hoots of chimpanzees across the water. In the chimps' night enclosure, Kidogo and Connie are faced with a dilemma. A wooden plank just out of reach is piled high with food on either end. To bring it within reach, they each have to pull on a rope threaded through metal loops on the plank. If only one of them pulls, the rope comes unthreaded and the plank stays where it is. Kidogo, a dominant female, pushes Connie out of the way and pulls on Connie's end of the rope—which then whizzes out of the loops so that no one gets any food.

This behavior is puzzling because chimpanzees in the wild are great cooperators, frequently hunting for food in what appears to be a complicated and organized fashion. But perhaps there is not much thinking going on behind this kind of cooperation; it could simply be that because each animal wants the same thing and all are at work at the same time, success happens by accident and just looks like a cooperative endeavor.

But if you watch Kidogo and Connie at feeding time, you will notice that they don't share food. If Connie has a piece of food and Kidogo is around, Kidogo will most likely steal it from her. On the other end of the spectrum, Sally and Becky have grown up together in the sanctuary and are like sisters. They share food peacefully, all the time. When you give them the rope test, they succeed on the first trial.

Clearly, if you allow for tolerance, chimpanzees can cooperate spontaneously. Not only do they know when they need someone, they also remember who's a good partner. Mawa, another dominant chimpanzee, is not a very good cooperator. He doesn't wait for his partner to pick up the other end of the rope and instead pulls it free of the plank. Bwambale, on the other hand, is a great cooperator; he waits for his partner, and they are nearly always successful in getting the food. At first, the other Ngamba chimps chose Mawa and Bwambale equally, but after Mawa botches it, most chimps chose Bwambale on the next trial.

However, such cooperation in chimpanzees is highly

constrained. Chimpanzees will cooperate only with familiar group members, with whom they normally share food. If they don't know or like a potential partner, they won't cooperate no matter how much food is at stake. Humans, however, make a living collaborating, even when it's with people they don't know and in many cases don't particularly like. (Do you have a boss?) This high level of social tolerance is likely one of the building blocks of the unique forms of cooperation seen in humans.

So perhaps a lack of tolerance is one of the main constraints on chimpanzees' developing more flexible cooperative skills. But humans have another closest relative, one who is usually forgotten and may be more like us than we know.

Long-Lost Cousins

In contrast to chimpanzees, who live in male-dominated societies with infanticidal tendencies and other forms of lethal aggression, bonobos live in societies that are highly tolerant and peaceful thanks to female dominance, which maintains group cohesion and regulates tensions through sexual behavior.

Since bonobos are more tolerant than chimpanzees, what does this mean for their cooperative abilities?

Further tests were done with the chimpanzees at Ngamba Island. As long as the food was in two separate piles on either end of the plank, most of the chimps could cooperate fine. But as soon as you put the food in one monopolizable pile in the middle, chimpanzee cooperation fell apart. Even though chimpanzees participating in the test were relatively tolerant of each other and had passed the rope/plank test many times before, whenever the food could be monopolized by a dominant chimp, the other chimp generally refused to pull.

When we gave the same test to bonobos, they played and had sex to negotiate with each other—even though this was their first run-through. Bonobos are notorious for their sexuality. Females rub their clitorises together; males have sexual activity with males. Neither age nor gender seems to matter. Sex is a tension-relieving activity in the group, used to soothe ruffled tempers or form alliances. It also appears to be a negotiating activity, engendering a high level of tolerance in bonobos.

So what we have are chimps who cooperate but aren't very tolerant, and bonobos who are very tolerant but don't really cooperate in the wild. What probably happened six million years ago, when hominids split from the ancestor we share with chimpanzees and bonobos, is that we became very tolerant, and this allowed us to cooperate in entirely new ways. Without this heightened tolerance, we would not be the species we are today.

Finding Our Minds in Africa

Spontaneous cooperation is not the only way in which bonobos are more like humans than chimpanzees are. As with humans, gender differences in bonobos are less pronounced. The males are not physically very different from the females. Female bonobos, like human females, develop strong bonds, whereas female chimps generally don't. Humans and bonobos have similar temperaments, in that we are both risk averse and wary of the new.

Understanding bonobos is crucial to understanding what

makes us human. Unfortunately, their numbers are dwindling fast. The only country where they're indigenous is the Democratic Republic of the Congo, and the various wars that periodically break out there have made studying them difficult. Africa's ape sanctuaries, including Lola Ya Bonobo, Ngamba Island, and the Tchimpounga Sanctuary for chimpanzees in the Republic of the Congo, offer an exciting opportunity to probe the minds of our closest relatives. Unlike lab animals, who are likely to suffer chronic psychological and physical problems in captivity, sanctuary apes live in large social groups in vast areas of tropical rain forest. The semicaptive apes can be tested in indoor enclosures, similar to conventional laboratories but much less costly. Sanctuary animals show no aberrant behavior (e.g., rocking or feces eating), and preliminary data suggest they may outperform captive apes in a variety of physical tasks, presumably because of the richness of their everyday environment.

Mikeno, the bonobo who sat like a Rodin sculpture, died in September 2006. An autopsy revealed a contusion on his brain, which suggests he died of a concussion after falling from a tree. Mikeno's close friend Isiro sat by him and refused to leave the body. Did she understand death? Did she feel a humanlike grief?

We're still a long way from discovering exactly what makes us human, but even if we do, there will still undoubtedly be a thousand more questions to answer about what makes a chimpanzee a chimpanzee and a bonobo a bonobo.

NATHAN WOLFE

holds the Lorry I. Lokey Visiting Professorship in Human Biology at Stanford University. He received his doctorate in immunology and infectious diseases from Harvard University in 1998, and has been the recipient of a Fulbright fellowship (1997), the NIH Director's Pioneer Award (2005), and the National Geographic Emerging Explorer award (2009). Wolfe's research aims to chart the diversity of microbial life on earth, and combines methods from molecular virology, ecology, evolutionary biology, and anthropology. Among his major findings is the first evidence of natural transmission of retroviruses from nonhuman primates to humans. He founded and directs the Global Viral Forecasting Initiative (GVFI), a pandemic early warning system that monitors the spillover of novel infectious agents from animals into humans. GVFI was recently awarded eleven million dollars from Google.org and the Skoll Foundation. It coordinates activities of over one hundred scientists and staff from countries around the world and currently has active research and public health projects in Cameroon, China, the Central African Republic, the Democratic Republic of the Congo, the Republic of the Congo, Gabon, Equatorial Guinea, Laos, Madagascar, Malaysia, and São Tomé.

THE ALIENS AMONG US

NATHAN WOLFE

Imagine yourself sitting at home on a comfortable couch, a cup of coffee in your hand, a book on your lap, the sound of your children playing in another room. Everything seems normal—as it has always been. But then again, maybe not. Perhaps, very vaguely, you sense that you are not alone.

Then quite suddenly the veil is lifted. Everywhere around you a formerly invisible life-form swarms—on the couch, coating the coffee cup, covering the pages of the book on your lap. An unrecognizable creature, small yet clearly alive, but with no cells, no enzymes, and a distinct absence of much of the biological machinery of the living beings with which we are familiar. This alien is everywhere; it has integrated itself, quite literally, into the fabric of life that surrounds us. Into every species of bacterium, every plant, fungus, and animal that makes up our world. It has invaded—and perhaps it has already won. It is . . . *the Virus*!

The fear of aliens haunts humanity—our movies, our books, our children's dreams. It drives our curiosity about the solar system. (Was there ever water on Mars? Could Mars have supported life?) It drives our curiosity about planets in distant solar systems. (Do they have atmospheres? Is oxygen present?) And it propels our exploration of the universe around us, pushing us to develop huge arrays of radio telescopes that search the universe for the distant cries of intelligent life.

Yet there is a certain irony in our search for aliens. While seeking far and wide, we largely ignore a particularly important planet. A planet with all the ingredients to support life. A planet we can explore with relative ease—no need for space suits or rockets. A planet that we already know supports life and is therefore arguably the most likely to support unknown life-forms. It is our Earth, and when it comes to the search for aliens in our universe, we are sadly neglecting it.

Already we know that Earth supports multiple kinds of life. It supports cellular life (the bacteria and their lately recognized and similar-looking cousins the archaea) as well as the eukaryotes—fungi, plants, and animals. It also supports what may arguably be another, rather unusual life-form, the prion, whose discovery was acknowledged with the award of a Nobel Prize in 1997. Prions are an odd form of life that lacks not only cells but also DNA or RNA, the genetic material that all other known forms of life on Earth use as their blueprint. Yet prions persist and seemingly replicate, causing, among other things, mad cow disease.

And our planet also supports viruses—nanocreatures with no cells that have penetrated the other forms of life on our planet with startling efficiency. Viruses are purely genetic material in a protein coat, and they are dependent on host cells, without which they can neither grow nor reproduce.

Unlike much of what's been written about viruses, this essay will not focus on the harm that viruses do to humans but rather take a forty-thousand-foot view of the diversity and abundance of viruses on the planet, their ecological significance, and their important role in enhancing the richness of life among the cell-based life-forms. It argues that a deeper understanding of viruses offers us new insights—not only into human health and disease but also into the basic biology of our planet. Furthermore, since viruses are such a profoundly alien form of life, understanding them in their full glory may give us a clue as to what encountering a real alien might be like.

They're Everywhere

Earth is a living planet. Even seen from space, it shows abundant evidence of life. Yet what we can see from space is only part of the story. The dominant forms of life on our planet, when measured in biomass and diversity, are microscopic. Microscopic life-forms are everywhere: in our oceans, on the land, deep underground. Deep-sea exploration using drills that dig hundreds of meters below the ocean floor reveals the presence of these organisms, which in this case live not off the energy of the sun but off the heat generated by Earth's core. And while the cellular microbes, such as bacteria and archaea, are perhaps easier to detect, acellular microbes, in the form of viruses, represent a vital part of this diversity.

It is thought that every form of cellular life hosts at least one type of virus. Every alga, bacterium, plant, insect, mammal. Everything. Even if every species of cellular life harbored only a single, unique virus, that would make viruses the most diverse known life-form on the planet. And many of these species, including humans, harbor a range of viruses.

Although viruses are small and light—the largest known is the still diminutive six-hundred-nanometer amoeba-infecting mimivirus—their incredible abundance leaves a sizable biological footprint. In a landmark 1989 paper, Oivind Bergh and colleagues at the University of Bergen, using electron microscopy to count viruses infecting bacteria in a range of unpolluted natural water ecosystems, reported up to 250 million virus particles per milliliter. Other more comprehensive estimates of the biomass of bacterial viruses in Earth's ecosystems stagger the imagination. One estimate suggests that if they were lined up head to tail, the resulting chain would extend 200 million light-years, far beyond the edge of our Milky Way galaxy (which measures only approximately 150 light-years across), with a mass estimated to be equivalent to approximately a million blue whales.[1] Far from a minor blight, viruses blanket our planet, with a richness and impact we are just beginning to understand.

Friend or Foe?

If you're like most people, when you think of a virus a specific idea comes to mind: a microscopic source of disease. Perhaps, if offered the opportunity to flip a switch and permanently

[1] N. H. Acheson, *Fundamentals of Molecular Virology* (New York: Wiley, 2006), 4.

eliminate all the viruses on the planet, you might choose to
flip the switch. But choose carefully. Could all those viruses,
inhabiting perhaps every cellular organism on the planet and
existing in such phenomenal diversity and abundance, be exclusively negative? The answer is a resounding No!

While viruses have to infect cellular forms of life in order to
complete their life cycles, this does not mean that causing dev-
astation is their destiny. The existing equilibrium of our planet
is dependent on the actions of the viral world, and their elimination would have profound consequences. It is estimated that
20 to 40 percent of bacteria in marine systems are killed by
viruses each day. These deaths provide a tremendous source of
organic matter, releasing amino acids, carbon, and nitrogen
into the marine environment and making viruses prime players
in the cycling of marine nutrients.[2] Moreover, although studies
of the role of viruses in regulating biological diversity are still in
their infancy, it is thought that viruses help to maintain the
diversity of key environmental players such as bacteria by acting as "trust busters"—that is, preventing any one bacterial
species from dominating.

But while nutrient cycling and the maintenance of diversity
are central roles, these are, of course, mediated by the viral role
in destroying living cells. Do viruses have the potential to benefit individual organisms? Unfortunately, much of the study of
human viruses has focused on viruses as agents of disease, and
this has negatively biased both our perspective on viruses and

[2] M. Middelboe and N. O. G. Jorgensen, "Viral Lysis of Bacteria: An Important Source of Dissolved Amino Acids and Cell Wall Compounds," *Journal of
the Marine Biological Association* 86 (2006): 605–12.

the research that has been undertaken to understand them. Thus it is not surprising that we're familiar mostly with viruses that cause harm. There have been no systematic attempts to identify beneficial viruses. What might we find if we looked for "good" viruses?

Viruses operate along a continuum with their hosts and other organisms they interact with: some harm their hosts, some benefit their hosts, and some—perhaps most—live in relative neutrality with them, neither substantively harming nor benefiting the organisms they must at least temporarily inhabit for their own survival. Although there are not many case studies of beneficial viruses, a fortuitous finding—the complicated relationship between a wasp, a caterpillar, and a virus—has shed light on the potential for viruses to aid their hosts. We normally think of wasps as colonial organisms living in nests, but there are a range of wasp species that nurture their young outside of group nests. One such noncolonial wasp family, the Braconidae, lays its eggs inside caterpillars. When the wasp eggs hatch, the larvae consume the caterpillar as their food. Needless to say, this comes at a substantial cost to caterpillars, and the result is an evolutionary arms race, with the caterpillars developing defenses against wasp egg-laying and the wasps developing countermeasures. Among these counter measures is a fascinating example of virus-host cooperation involving a polydnavirus, a type of virus that uses DNA, rather than RNA, as its genetic information. This virus has evolved a mutually beneficial relationship with the wasp. It replicates in the wasp's ovaries and is injected, together with the wasp's eggs, into the caterpillar; the virus returns the favor by suppressing the host caterpillar's immune system and thereby protecting the eggs. The wasp helps the virus, and the virus helps the wasp.[3]

Given the ubiquity of viruses, it would be surprising indeed if they were simply relegated to a destructive role. Further studies will likely reveal the profound ecological importance of these organisms not just in destroying but also in benefiting many of the organisms they infect.

Instigators of Evolution

Among the benefits that viruses can provide is one at the heart of evolution: genetic diversity. Because of their high mutation rates and their ability to exchange genetic information with one another, viruses are tremendous generators of genetic variation. They constantly move genetic information from organism to organism, occasionally introducing novel genetic information from themselves and their prior hosts into new hosts. This genetic diversity can act as fuel in the process of evolution, providing new genes that natural selection can act on. While the genes moved around by viruses are not always beneficial to their new recipients, every so often they provide pivotal advantages. Classic examples include the toxins of some bacterial species, such as cholera, which are derived from the genes of viruses. Virulent strains of cholera carry a virus called CTXf, which has the genes for a toxin that causes massive diarrhea — thereby helping the cholera bacteria to spread quickly.

[3] K. M. Edson et al., "Virus in a Parasitoid Wasp: Suppression of the Cellular Immune Response in the Parasitoid's Host," *Science* 211(1981): 582–83.

But such genetic introduction is by no means limited to bacteria. Retroviruses, which integrate their genetic material into the DNA of their hosts, can do the same for animals. It now appears that the introduction of a retrovirus gene into our ape ancestors led to a new mammalian gene that plays an important role in our placenta.[4] The gene, ERVWE1, clearly derives from a retrovirus and has now become a proper and permanent gene in humans. Given that retroviruses and fetuses alike have to negotiate the complexities of the immune system, and the immune system's constant efforts to reject the "other," it is perhaps not surprising that the genes of retroviruses could be of use in our reproductive system. It is thought that ERVWE1 has precisely this function—that is, it helps in suppressing the immune system so as to avoid rejection of the "foreign" fetus, just as the retrovirus must control the immune system to avoid its own rejection. Through the generation and ferrying of genetic diversity, viruses provide vital creative fuel and fluidity to the evolution of life on Earth.

Vindicating Viruses

The word "virus" comes from the Latin *virus*, which means "toxin" or "poison." And while our understanding of the importance of viruses for life on the planet has changed substantially since the word itself first entered the English language (in the fourteenth century, as a synonym for "venomous substance"), the science of viruses continues to reflect those early negative

[4] F. Mallet et al., "The Endogenous Retroviral Locus ERVWE1 Is a Bona Fide Gene Involved in Hominoid Placental Physiology," *Proceedings of the National Academy of Sciences* 1(2004): 1731–36.

misconceptions. Some virologists have begun to embrace a more holistic view of the viral role. Yet the vast majority of the public, as well as many scientists who study the viruses of vertebrates, still focus almost exclusively on identifying and responding to those that harm human and animal populations.

Misconceptions of the importance of viruses have also affected the disciplines in which viruses are studied. Viral sciences are seen as subdisciplines of microbiology and human medicine, and this must be corrected. The study of viruses must be brought forward and expanded as a primary objective of science. Initiatives to characterize and understand global viral diversity promise a range of immediate and long-term benefits, from preventing the next major pandemic to understanding carbon cycles in the ocean. A new generation of thinkers in evolutionary biology, physics, earth sciences, and computer science should converge with physicians and molecular biologists to understand what increasingly appears to be one of the most important forms of life on our planet.

Fortunately, there are new tools that place the dawn of viral science on the horizon. Metagenomics, which involves using genetic-sequencing techniques to characterize the diversity of genomes in specimens from biologically rich environments, ranging from soil to skin, has begun to reveal the richness of viral and other microbial life in our world. Studies with human feces, for example, suggest the presence of more than a thousand distinct viruses in a given stool sample—more viruses than some once thought infected the entire human species![5] Clearly our knowledge of the viruses that

[5] M. Breitbart et al., "Metagenomic Analyses of an Uncultured Viral Community from Human Feces," *Journal of Bacteriology* 185(2003): 6220–23.

inhabit us and our world is in its infancy. Further such studies, and related studies of animals, plants, soils, and aquatic systems, represent a new voyage of discovery. The viral collections that result will constitute a twenty-first-century viral museum of use to scientists in understanding the environment, forecasting the next pandemic, developing new antibiotics, and discovering and introducing new genes for a range of useful purposes.

A new, holistic viral science has enormous potential. Viruses have already proved invaluable as vaccines and as tools in molecular biology. The use of unadulterated vaccinia virus, a variant of the cowpox virus, allowed humans to wipe smallpox, perhaps the worst disease scourge humanity has ever faced, off the face of the Earth. And bacteriophages (bacterial viruses) have proved to be important models for basic discoveries in molecular biology, as well as molecular tools (vectors) for the introduction of new genes into living cells—most spectacularly in the recent successful creation of nonembryonic stem cells.[6] It is impossible to imagine the many discoveries of basic and practical relevance that await us as this research unfolds.

Beyond the practical significance of comprehensive global viral exploration, the study of viruses also has something to tell us about our own place in the universe. Viruses represent a fundamentally distinct and simpler form of life—one that may actually be more common in the universe than more complex forms.[7] They may well tell us what it might be like to discover

[6] J. Yu, "Induced Pluripotent Stem Cell Lines Derived from Human Somatic Cells," *Science* 318(2007): 1917–20.

[7] C. H. Lineweaver, "We Have Not Detected Extraterrestrial Life, or Have We?" in *Life As We Know It*, J. Seckbach, ed. (Dordrecht, Netherlands: Springer, 2006), 445–57.

and interact with extraterrestrials. Simply finding viruses has proved to be a difficult task, suggesting that we would be wrong to assume that a "first contact" would be noticed at all. We've barely scratched the surface of the viral world, which should make us wonder about the incredible difficulty of identifying real aliens.

The aliens of popular imagination are usually portrayed as either perfectly benign or perfectly evil. Our stereotyping of viruses is similar; in reality, they represent the full spectrum of good, bad, ugly, and simply benign. It is almost certainly wrong to assume that extraterrestrial life will be any different. Moreover, the question of whether or not aliens will be aware of us seems based on a particularly anthropocentric view of the universe. It may well be that first contact will be a dramatic encounter, either with gracious beings of higher intelligence or evil warlords from another planet. But it is much more likely that they will prove hard to find, even if they come to inhabit the space around us—neither extraordinarily good nor extraordinarily bad for us as a species, and of enormous benefit to our science.[8]

[8] Douglas Morier assisted in the background research for this essay.

SEIRIAN SUMNER

is a research fellow in evolutionary biology at the Institute of Zoology, Zoological Society of London. She has a B.S. in zoology (1995) and a Ph.D. in behavioral ecology and evolution (1999) from University College London and worked at the University of Copenhagen and the Smithsonian Tropical Research Institute in Panama before settling at the Institute of Zoology in 2004. She has held a succession of research fellowships, including most recently the L'Oréal for Women in Science Fellowship.

Sumner's research focuses on the evolution of sociality—how eusociality evolves and how social behavior is maintained. She has worked with a variety of bees, wasps, and ants from around the world, studying their behavior through observation, experimental manipulation, and molecular analyses, including gene expression. She is especially interested in the origins of sociality and the role of the genome in this major evolutionary transition.

HOW DID THE SOCIAL INSECTS BECOME SOCIAL?

SEIRIAN SUMNER

You need to move to accommodate your growing family. You would like to stay in the area to keep your moving costs down and because you really like the local grocery store. After a bit of searching, you and your family decide on a new home. Under your guidance, the family works well together as a team: the youngsters help pack up the old house, the older ones carry the luggage to the new house. The girls do the bulk of the work, so it is fortunate that you have a lot of daughters. Girls are also good at organizing, so the new house is soon up and running. Your youngest is already excitedly telling her siblings about the candy store she's discovered down the road.

This could be the story of any family anywhere in the world: this one is about a family of ants. The similarities between the lives of social insects, like ants, and our own human societies are astonishing and reach far beyond house-moving analogy. Waste disposal, undertaking, child care, monarchies, slave trade, free riders, punishment, rebellions, architecture, agriculture, murder, cannibalism—social insects have it all, and they did it all eighty million years before we did.

Our fascination with social-insect societies and how they balance conflict and cooperation is therefore hardly surprising. For well over a century, scientists have described them, watched them, measured them, and even squashed them up to look at their genes. We are now equipped with a comprehensive body of scientific literature on how these enormous empires evolve and are maintained. We know that these animals are brilliant communicators: honeybees describe to one another the location of food sources in their famous waggle dance, rivaling any car's satellite navigation system. They are tolerant mutualists: the obligate alliance between leaf-cutting ants and their fungus gardens makes the most loved-up Hollywood couples look coolly unacquainted. They are also land conquerors: a single family of the Argentine ant *Linepithema humile* extends across six thousand kilometers, far exceeding that of any human conurbation, let alone a single human family. The biomass of ants alone accounts for 20 percent of the world's animal biomass and far exceeds that of all living humans. There is no doubt that the eusocial insects— i.e., those whose colonies exhibit overlapping generations, reproductive division of labor, and cooperative brood care—are a natural marvel and deserving of our full attention.

The secret to the success of eusocial insects lies in their task partitioning. Individual colony members specialize in either reproduction (the queen caste and males) or foraging and brood care (the worker caste, all of which are females). In honeybee colonies, as with most eusocial insects, workers have lost the ability to mate and so cannot lay fertilized (female) eggs. Instead, the colony puts all its genes in one basket, and that basket is the queen. Once this adaptation is achieved, castes are well on their way down evolution's slippery slope into the oblivion of caste permanency. Evolution has handed workers a rough deal, since they do all the work but never get to have sex. This was one of Charles Darwin's biggest headaches when he was devising his theory of natural selection. He surely rests easy in his grave, now that we know that evolving permanent sterile castes was undeniably the cleverest thing social insects could do. Castes reduce conflict: if you cannot pump out the eggs, you are not going to waste time and energy challenging your reproductive leader. Less conflict computes into more of your genes being passed on, in the form of your relative's offspring (usually your siblings). By limiting the number of queens in a colony, workers can be more certain of their relatedness to the brood they help raise—and consequently of their genetic payback from working. Once a despotic social structure emerges, colonies can grow, with millions of workers serving a single queen.

As we slither farther down evolution's helter-skelter slope of eusocial complexity, the worker caste divides elegantly into poetic polymorphisms: major, median, minor, and minim workers carry out different tasks, which are determined by their size, making the gene-propagating conveyor belt of colony life run all the more smoothly. Finally, at what we think is the limit of social complexity, a third caste emerges: large and ferocious, these are the soldiers, who are equipped for colony defense with their mandibular fangs and stinging spears, which have inspired many a science-fiction writer—fiction based on fact.

These highly complex societies are awe inspiring and have understandably won the lion's share of attention from sociobiologists. But they can tell us little about the *origins* of eusociality, since such societies are evolutionarily derived, making it difficult to tease cause from consequence. The ultimate question for the next generation of sociobiologists is: what is the secret to *becoming* eusocial? We would like to know what the conditions and selection pressures were that tipped the ancestors of the eusocial insects over the ledge and down toward eusociality. Crucially, how did the first castes evolve?

Let us start at the very beginning. The ancestors of social insects were solitary, and living representatives are readily found in nature. A single digger wasp will lay her egg in an excavated hole in the ground, provision it with a juicy caterpillar, and leave her young to feed from the larder without ever knowing their parents (surely every adolescent's dream). Next up, we have the simplest insect societies, which consist of two or more females cohabiting in a nest and sharing tasks, such as nest building, nest guarding, and foraging. If this proves more successful than nesting alone, we have taken the first step toward eusociality. The ultimate breakthrough in social evolution therefore is transiting from a solitary life, such as that of the lonesome digger wasp, to a social one—be it a small, egalitarian society of totipotent females or a massive, highly structured empire with a million laboring subjects ruled by a single queen. With only one in fifty insect species being eusocial, the message is clear: transiting from solitary to eusocial life is difficult. But why is it so difficult?

The natural world is full of brilliant mathematicians (animals, plants, fungi, algae), who are expertly balancing the relative costs and benefits of different life strategies on the evolutionary scales of genetic fitness. The late evolutionary biologist

W. D. Hamilton pointed out that social insects have to be a bit cleverer than your average animal, as they need to assess how many genes they share with nestmates. Colony members will accept a seemingly unfair division of reproductive labor if their dominant nestmates are close relatives. In verbal terms, Hamilton's rule is this: if the benefits (b) of helping raise a relative's brood, weighted by your relatedness to these brood (r), exceed the costs (c) you incur in sacrificing your own chance of breeding, it pays to be social—that is, be social when $br > c$. Few scientific breakthroughs are simpler or more elegant than Hamilton's rule, and it has served us well in our quest to understand eusocial evolution. It cannot, however, explain how that first step onto the ladder of eusociality was achieved. How did the conditions required by Hamilton's rule first come about? How did females find relatives and decide to nest cooperatively?

The holy grail of the ultimate breakthrough lies locked inside socially polymorphic bees and wasps. They could be the missing link between a solitary and a eusocial way of life. These species can choose to be social or solitary, depending on where they live, when they were born, who their neighbors are, or what the weather is like. If we can understand how these decisions are made, then we are starting to acquaint ourselves with eusociality's ancestor. Come with me into the world of a socially polymorphic bee. What are her dilemmas, and how does she solve them?

Dilemma 1: Whether to Be Social

Consider our bee, a female, emerging from hibernation into the gentle breeze of an English spring. Can she assess the merits of being social rather than solitary? The decision is not unlike finding an apartment to rent: you could choose to rent with other people, which will be cheaper and more efficient (you can share housework, shopping, etc.), but you have to put up with the inconvenience of other people's mess and antisocial behavior. The alternative—renting alone—will be more expensive and risky: you will need to do all the chores, and what if your house is burgled while you're out shopping? On the plus side, at least you have peace, privacy, and, most important, control. When we make these decisions, we are weighing the costs and benefits of alternative strategies, just like our emerging bee.

If our bee has emerged in the north of Scotland, she may have a frosty reception into a cool and short-lived summer. She would do well to nest alone, raising a single brood, who will become next year's queens. In the south of England, where summers are longer and warmer, the same species of bee is found nesting alone, just like its northern counterparts, but it is also found nesting eusocially. In these eusocial colonies, the first brood laid will behave as workers and help raise the second lot of brood, who will be next year's queens. The decision about whether to be social or solitary is evidently influenced by the environment.

At school I hung out with pretty much the same group of girls for five years. The group hierarchy was carefully monitored by a self-appointed queen, who exercised her power by rotating "best friend" status among us, depending on her mood. It was an excellent divide-and-rule strategy, which kept everyone vulnerable, hopeful, and occasionally revered. Hanging out in a group was a badge of popularity and guaranteed a supply of boys. A good reason for our bee to become social is if she thinks she can be queen, or at least a favored subordinate ("best friend"). Some colonies may be egalitarian, in which case being a worker does not really mean much, but others could be despotic. On the cusp of becoming eusocial, our bee needs to know what her chances are of becoming an egg-layer if she plunges into a eusocial life. Swarms of theoretical models produced over the last fifteen years have attempted to explain how reproductive partitioning is decided. According to these models, instead of handing out popularity to ensure group cohesion, as my school queen did, social-insect queens hand out breeding incentives to group members depending on social, environmental, and physical factors. Just like my school queen, a social-insect queen *needs* her group, and so she may persuade other females to stay and help her by offering them some share of egg-laying. Social contracts like this appear to work for paper wasps: the dominant female will allow her subordinates to lay some eggs, just to keep them happy. The females of other eusocial species, such as allodapine bees, compete with each other for the better share of reproduction.

Whatever the strategy, social insects in the simplest societies are able to (indirectly) weigh the relative costs and benefits of different strategies. We are slowly accumulating data on how reproduction is divided in simple societies where all females are potential queens, but we do not seem to get any closer to an answer. Every new species studied seems to support a different theoretical model. Maybe it is time we gave up on the quest for a universal rule. Or perhaps we are looking at things from the wrong angle; a population- or species-level approach might be more informative.

Dilemma 2: How to Be Social

Once she has decided to be social, our bee needs to find a suitable nestmate. Where does she find one, and how can she be sure that the candidate is a close relative? The implications of getting it wrong—choosing to nest with an unrelated female—are catastrophic, as workers rely on propagating their genes by helping raise a related brood. Recognizing a relative could make use of simple rules like "Nest with someone who emerges from hibernation near you," since your sisters are likely to hibernate close to where you emerge. Or "Nest with someone who looks a bit like you," choosing by means of a defining characteristic, such as Richard Dawkins's "green beard" label. In *The Selfish Gene*, he notes that such unique characteristics are "one way a gene might 'recognize' copies of itself in other individuals."

I have a friend who shares my surname: I'm tempted to use this as an indication that we're related; my surname is not very common, and she does look a bit like me (no, we don't both have green beards). However, she is from a different country. Unless we do a genetic test, I won't be able to tell whether she's a closer relative than the next person I pass on the street. Do social insects have superior powers of detecting kinship? A clever solution, adopted by the simplest eusocial animals, is to set up camp alone and let your offspring be your workers—but how can they be sure that you are their mom and not a social parasite, an interloper masquerading as your mom?

When the white-coated chemists invited the welly-shod sociobiologists into the chemistry lab about twenty years ago, some headway was made on working out the mechanisms used by social insects in recognizing kin. Perfume and makeup, that's all it takes. Hydrocarbons on the insect's surface enable individuals to recognize nestmates. Scientists have had lots of fun dousing wasps in the "wrong" nest perfume and watching nestmates recoil in horror at the presence of a perceived intruder. Chemicals are not foolproof, though. Socially parasitic wasps have an inert scent, which allows them to invade a host nest unnoticed. Over time, they acquire the nest odor, and host workers start working for the intruders because they are unable to distinguish them from real nestmates. Other scientists have applied a little eyeliner and rouge to the faces of paper wasps, which can really mess up the wasp recognition system.

But are not smelliness and looks learned recognition cues, rather than true indicators of an insect's innate skill in balancing relatedness coefficients? All that scientists are certain of is that social insects can indeed distinguish nestmates from nonnestmates, just as you know who your friends and neighbors are. Social insects may not, however, be the superbeings we imagine they are, comparing coefficients of relatedness for individual brood or adults within their nests. There are suggestions that highly derived species, like some ants, can do this, but the definitive experiment has yet to be carried out. Of course, even the cleverest insects make mistakes, but we don't like to talk about that much. Honeybees, bumblebees, and vespine and polistine wasps are frequently found working in the "wrong" nest. In fact, the more we look, the more we find these "errors." Let us do Hamilton a favor by looking outside the box and embracing new (kin-selection) explanations for these "mistakes."

Dilemma 3: When to Be Social

Our British bee has reason to drone on about the awful British weather. After all, it influences whether she can be social or not. For her, the decision on whether to be social has to be made early in the spring. Perhaps her rule goes something like this: "If it doesn't get above 10°C by mid-April, best to be solitary, as it is going to be a chilly/short breeding season—no time for two broods." If there is a rule, how often is it broken? Have the solitary females in the warmer, southern Britain just made mistakes?

I have done a lot of my research in Panama, where I enjoy a balanced, less hectic approach to life. Our time-stressed, sun-deprived British bee deserves a holiday in Panama: she could chill out a bit and take life as it comes. She might set up house alone initially. When her first brood emerged, she could decide whether to kick them out and resume her solitude or become eusocial and entice them to stay. Or perhaps she'll eventually tire of her solitude and strike up a coalition with a related neighbor. Unlike life as a bee in a temperate latitude, it probably pays a subtropical bee to retain behavioral plasticity throughout her life so that she can adopt different strategies opportunistically in an aseasonal environment. It may be that permanent morphological castes evolved in temperate regions.

We need to know what happens if our bee makes the wrong decision. How flexible is she? Many simple societies of obligate eusocial bees and wasps can switch from worker to queen, such that worker behavior is a temporary option for females while they wait for a better option, either in their own nest or in one nearby. What if our bee's coalition fails? Can she retrench as a solitary female? For a female with high behavioral plasticity, jumping ship to a nearby nest with sunnier reproductive prospects is a viable option.

At what point did eusocial insects lose this plasticity? Some species are age limited in caste plasticity. Is this the next step toward evolving permanent castes? Or is ecology the limiting factor? There might not be any single-occupant houses left to nest in. If we can figure out the genetic, environmental, and ecological factors that govern a caste's behavioral flexibility and work out when plasticity is lost, the fog will start to clear.

Evolving Eusociality at the Ultimate Level

Despite decades of obsession with highly derived, complex eusocial species such as honeybees and ants rather than the simpler insect societies, we actually know quite a lot about the origins of sociality. We know a bit about the mechanisms of how our bee becomes social, when and why she does it, and how she deals with the burden of constantly assessing Hamilton's rule. But it is time to take our story on the origins of eusociality and strip it down to its most fundamental state.

Biologists, like it or not, are now pawns in an "omics" world: genomics, proteomics, and transcriptomics describe, respectively, the study of genes, proteins, and transcripts (in which genes are expressed). Anything can be "omics"ed: metabolomics, interactomics (oh, behave!). "Sociomics" is up there, too—the study of the genes and molecules that make up social systems (see http://omics.org). Marvelous! The Dawn of the Genome Giants looms over biology, stripping the classical elegance of the natural world down to the bare genes, making those musty mysteries exposed and vulnerable. A peek inside the genomics of our bee could allow us to discover the ancestors of the eusocial insects. By looking at how "solitary" and "social" genes are expressed at different stages of her life cycle, we might finally get close enough to discover whether those ancestors were closet socialites locked in a solitary world or hermits who were only reluctantly social. Or perhaps they weren't either; perhaps they evolved to be flexible, to take the opportunities of life as they come. If we can tackle this question, we are truly homing in on eusociality's ancestors and the ultimate breakthrough.

Thanks to the Genome Giants and their work on model organisms like *Drosophila* (the fruit fly), *Caenorhabditis elegans* (the nematode worm), and more recently *Apis mellifera* (the honeybee), we can make a start on studying the genes underlying the origins of eusociality, taking our understanding to a whole new level. Moreover, new questions we have never dared think about can now be addressed. For example, how does the genome change during the transition to eusocial living? Are certain genes copied, reshuffled, mutated? If so, which ones? What sort of "solitary" genes might be lost? Do new "social" genes with novel functions evolve? When should the gene plasticity present at the dawn of eusociality be sacrificed for rigidity, reliability, and permanence?

I hope I have now convinced you that the secrets of eusocial insects are rich and entertaining, and that we are only starting to understand how they originated. Given the analogies that can be drawn between their lives and ours, and the extra eighty million years of evolution that social insects have endured, they are undoubtedly wise as well as wonderful. As the self-appointed superiors of the animal kingdom, we stand humble before social-insect colonies.

I began this essay with a list of analogies comparing the parallel universes of insect and human societies, suggesting there is much we can learn from them, a clear example being the division of labor among group members. By some measures it would appear that we are not the best of pupils: for example, like many women in today's go-getting human communities, I juggle breeding (my "queen" side?) with work, foraging in the hungry world of science. Humans would do well to heed the lesson from social insects (which, after all, have had a few million years' head start on us) by hastening the emerging shift to a skills-determined division of labor from a gender-determined one. But that's another essay altogether.

KATERINA HARVATI

is a paleoanthropologist specializing in Neanderthal evolution and modern human origins. Her research interests include evolutionary theory, the relationship between morphological variation and genetic and environmental factors, and the evolution of primate and human life history. She conducts fieldwork in Africa and Europe, including her native Greece.

Harvati studied at Columbia University, the City University of New York, and the American Museum of Natural History. Before joining the Max Planck Institute for Evolutionary Anthropology in 2004, she was an assistant professor in the Department of Anthropology at New York University. She is also an adjunct associate professor of anthropology at the City University of New York Graduate Center. Analysis by Harvati and her colleagues on a late Pleistocene human skull found in South Africa's Eastern Cape Province was listed by Time *magazine as one of the top ten scientific discoveries of 2007, constituting "the first fossil evidence that modern humans left Africa between 65,000 and 25,000 years ago."*[1]

[1] F. E. Grine, R. M. Bailey, K. Harvati, et al., "Late Pleistocene Human Skull from Hofmeyr, South Africa, and Modern Human Origins," *Science* 315, no. 5809(2007): 226–29.

EXTINCTION AND THE EVOLUTION OF HUMANKIND

KATERINA HARVATI

It is almost a cliché in the extinction literature that Charles Darwin and most subsequent prominent evolutionary thinkers did not pay much attention to extinction as a major force of evolution. Given that the vast majority of species that ever existed on Earth are no longer with us, this is somewhat surprising. The processes of extinction have been seriously investigated only in the last few decades—now that the magnitude of the human impact on the planet has become obvious and the biodiversity crisis stares us in the face.

Today even a cursory glance at the 2008 IUCN Red List of Threatened Species (http://www.iucnredlist.org) reveals the enormous extent of the current extinction crisis and how it might affect the prospects for our own survival. This impression is reinforced by an examination of the fossil record. Along with the normal hum of "background" extinctions—the loss of single species here and there—several mass-extinction events have punctuated the history of life. Some were severe enough to kill off as many as half the existing taxa and resulted in a complete restructuring of Earth's biomes. Perhaps the most widely recognized of these (though not the most severe) is the K/T (Cretaceous/Tertiary) event, some sixty-five million years ago, best known for the demise of the dinosaurs and the subsequent prolific evolution of mammals.

Although no such dramatic mass extinctions have occurred (perhaps until now) in our own lifetime as a species, extinction has nonetheless been important in shaping our evolution. Two relatively clear-cut extinction events can be discerned in the human fossil record: that of the genus *Paranthropus* about a million years ago and of the Neanderthals some thirty thousand years ago. Both extinctions were defining moments in the history of our lineage.

Why Do Taxa Go Extinct?

The answer to this question seems obvious and hardly worth considering, but the truth is that we don't really know. The causes of extinction are complex and not well understood. The direct or indirect effects of human activity are a critical factor in our current situation but were not important in the past. Human agency has been claimed, by some researchers, for a few extinction events in the very recent fossil record, most notably the megafaunal extinctions in Australia and the Americas about fifty thousand and eleven thousand years ago, respectively. But beyond the brief temporal reach of *Homo sapiens*, other important factors must be at work for species to die out and, even more important, for the plants and animals of an entire region to disappear.

There are two principal views on the causes of extinction. The first, favored by Darwin, focuses on competition with other taxa, often close relatives of the doomed species. Since natural selection produces improved varieties, new species are most often improved versions of their ancestors. When the two come into contact (as when landmasses reunite, bringing together closely related species that had previously lived in isolation from each other), the descendants generally outcompete the ancestors, causing the extinction of the older species. This view has led to the formulation of the principle of competitive exclusion, which holds that two species with the same ecological requirements cannot coexist for long in one region; one of the two will eventually drive the other to extinction, either because it is inherently more efficient at exploiting resources or, if there is no competitive advantage, because of its greater initial population. Although we often think of competition in terms of aggression, it occurs whenever two species need to obtain the same limited resource, such as food or nesting grounds, from the environment. Cases of invading species are extreme examples; such invasions are known from the fossil record—for example, the migration of North American taxa to South America when the two continents joined through the Isthmus of Panama about 3.5 million years ago. During this event, referred to as the Great American Biotic Interchange, much of the endemic South American fauna became extinct, replaced by taxa invading from the north. Human introduction of alien species has also produced extinctions of endemic faunas.

The alternative common explanation for extinction is environmental—usually climatic—change. This view sees extinction as a more-or-less chance event brought about not because of a species' inferior capability but because of bad luck; the species was, in a sense, in the wrong place at the wrong time. This type of explanation includes increased volcanism, changes in sea level, shifts in the position of the continental masses, and changes in Earth's orbital parameters, all of which can trigger climate change or habitat loss. Such geological events are common in Earth's history and therefore might be good explanations for the occurrence of background extinctions—but because of their frequency and sometimes relatively gradual nature, they are not completely satisfactory as causes of mass extinctions. It has been argued that sudden, catastrophic change is required if even a single species is to become extinct—that the environmental change must be rare enough to be unprecedented in the species' history and sudden enough that there is no time for natural selection to act to avoid the extinction. On the catastrophic end of the continuum of environmental phenomena that have been blamed for extinction events are rare and devastating natural disasters such as asteroid or comet impacts. Such an impact is strongly suspected in the K/T mass extinction, and similar catastrophes might be implicated in similar mass extinctions documented in the fossil record, though this picture may be influenced by the poor preservation of fossil-bearing sediments.

Not all species are equally vulnerable to extinction, whether caused by competition or environmental change. Although no particular attribute seems sufficient to save a taxon in the event of a mass extinction, a few characteristics offer some protection in less dire circumstances. A large geographic distribution is advantageous, as is large population size, high potential for dispersal, and the ability to tolerate and make use of a wide range of environments. Factors that increase vulnerability include large body size and the extended gestation and maturation periods with which it is often associated, dependence on a narrow adaptive niche, and, perhaps most important, small population size. Island taxa are particularly susceptible to swift extinction.

Not All Human Fossils Are Our Ancestors

The idea that human taxa could become extinct without issue has not always been widely accepted and still faces resistance in some paleoanthropological circles. Its twin and necessary companion is that two (or more) human taxa could have been alive at the same time in the past, a situation very different from our experience, which is based on the current status of *H. sapiens* as a single, widespread, and geographically variable species. Taking the current condition as the norm, evolutionary thinkers in the 1950s proposed the single-species hypothesis, developed from the principle of competitive exclusion as it would apply to human fossil taxa. This hypothesis posited that once human ancestors had developed tool use and culture, it became their adaptive niche, enabling them to exploit a wide range of environments and expand their territory. Because culture was the human adaptation, no two culture-bearing human taxa could exist at any one time in one place, according to competitive exclusion. Furthermore, because culture was thought to facilitate environmental tolerance and geographic spread, the isolation of a human lineage long enough for speciation was deemed virtually impossible.

This hypothesis was applied, almost without question, to the more recent Pleistocene fossil record and was later extended by some to the earlier australopithecines as well. It dispensed with both speciation and extinction in the human lineage. By assuming that there is only one human species at any given time, you are also assuming that it could not have gone extinct (at least not in the terminal, phyletic sense—that is, without evolving into a new species) because we are here! This approach put fossil humans in a peculiar situation, essentially removing them from the realm of the natural world and placing them squarely within the sheltered cocoon of what we perceive as modern humanity.

Paleoanthropology has come a long way since the single-species hypothesis was proposed. Advances in the field increasingly point to a bushlike pattern of human evolution, with multiple hominin species perhaps being the norm through most of our lineage. It is now clear that humans (whether fossil or living) are not immune from biological forces and that extinction was (and, indeed, is) a distinct possibility.

The Case of Paranthropus

Paranthropus (sometimes also referred to as robust australopithecines) appears in the human fossil record some 2.7 million years ago, in East and South Africa. Although it was bipedal like us, its brain was little different from those of apes. It possessed highly diagnostic morphology linked to powerful masticatory apparatus: flaring cheekbones, an almost concave face, sagittal crests on top of the cranium, and large cheek teeth and jaws. This extreme morphology indicated a dietary specialization and also placed *Paranthropus* beyond the acceptable range of anatomical variation of older *Australopithecus* taxa, dealing a decisive blow to the single-species hypothesis in the earlier part of human evolution.

Paranthropus is so divergent in its craniodental morphology that it forms a distinct genus, itself comprising as many as three species, all of which are too different from later hominins to have given rise to them. Although the fossil record is spotty and not always well dated, the last specimens of this genus date to about a million years before the present. The absolute extinction of *Paranthropus* is not in dispute—but what caused it? And what allowed the differentiation of this genus in the first place?

The origin of *Paranthropus* roughly coincided with the appearance of early representatives of our own genus, *Homo*, and of early stone tools in Africa, as well as with a climatic shift toward increased aridity. It is commonly thought that this climatic change brought about the evolution of the two lineages, *Paranthropus* and *Homo*, from a common australopith ancestor. Until recently, *Paranthropus* was thought to have adapted to the new conditions by evolving its extreme masticatory anatomy, while *Homo* followed a different route, widening its adaptive niche by developing stone-tool technology (hence culture). Because only one of the two genera, *Homo*, possessed stone-tool technology, the two hominins could coexist without violating the exclusion principle. This scenario assumes that *Homo* was the only maker of the early tools, but this assumption has been questioned. *Paranthropus* possessed the dexterity necessary to make tools and is thought to have produced bone-tool assemblages probably used for such activities as termite fishing.

Perhaps it was a narrow adaptive niche—in terms of diet—that allowed *Paranthropus* to coexist with a presumedly more generalized early *Homo*. Since specialist species are more vulnerable to extinction than generalist ones, and given the intensification of the glacial cycles around a million years ago, specialization could explain *Paranthropus*'s demise. Nonetheless, recent evidence suggests that despite its impressive masticatory adaptation, *Paranthropus* was no more specialized, either in its dietary or habitat preferences, than early *Homo*. In fact it seems that *Paranthropus* scores high on some of the criteria for invulnerability: it is by far the most abundant taxon in the fossil record in the period of its existence, which suggests a large population; and one of its species, *P. boisei*, is one of the longest-lived, and hence evolutionarily successful, hominins, spanning more than a million years.

The factors that led *Paranthropus* to extinction are not clear. They may have involved key environmental changes or competition with another taxon, possibly *Homo*. What is certain is that *Paranthropus* vanishes from the fossil record about a million years ago, leaving early *Homo* as the sole representative of our lineage.

A Sibling Species

Unlike *Paranthropus*, Neanderthals were very much like us: large-bodied, large-brained hominins, with sophisticated and well-documented cultural behavior. They lived in Ice Age Europe, from about three lakhs years ago to as recently as thirty thousand years ago. Like us, they were hunters and top predators, used fire, and buried their dead. But they were also sharply different: placed in a comparative primate framework, the anatomical difference between Neanderthal and modern human craniums is larger than the difference between that of the two species of chimpanzee. Nowadays Neanderthals are commonly considered a distinct species, *Homo neanderthalensis*. They were the descendants of a lineage that split from ours about half a million years ago, when it migrated from Africa and became isolated in Europe. Geographic isolation, then, probably accounts for the simultaneous evolution of two human species. Although half a million years seems like a long time, it is brief in evolutionary terms. Neanderthals were our closest relatives, much more closely related to us than our closest living relatives, the chimpanzees. They were our sister species.

Neanderthals disappeared from the fossil record around thirty thousand years ago, leaving no descendants and no detectable contribution to the modern human gene pool. Their extinction occurred a few thousand years after the arrival of modern humans (*H. sapiens*) in Europe some forty thousand years ago. So the obvious question is: did an environmental shift cause the disappearance of Neanderthals, as has been proposed for so many other extinct taxa? Or did the arrival of *H. sapiens* have something to do with it—perhaps something violent?

An abrupt, catastrophic climatic change is the easier scenario to rule out. No climatic event of great enough magnitude is associated with any of the proposed dates for the Neanderthal extinction.[2] Nonetheless, extreme climatic instability

[2] P. C. Tzedakis et al., "Placing Late Neanderthals in a Climatic Context," *Nature* 449(2007): 206–8.

may have stressed Neanderthal populations and brought them to a slow decline. But climate stress alone seems an unsatisfactory explanation; after all, Neanderthals evolved in Europe and had successfully survived previous intense climatic shifts.

However, climate stress probably set the stage for intensified competition with an invading closely related species. Can we document competition? There is no evidence of aggressive interaction between *H. sapiens* and *H. neanderthalensis*—but competition is much more than aggression. Modern humans hunted the same large herbivores that Neanderthals hunted. If this overlap in resource use was great enough, the argument for competitive exclusion might apply. Given enough time, one of the two might drive the other to extinction, even in a landmass as large as Western Eurasia.

Even so, Neanderthals had a few evolutionary advantages over our ancestors. They likely matured somewhat faster than early modern humans and thus would have taken less time to reproduce. They also had the home-court advantage: for hundreds of thousands of years, they had evolved in Ice Age Europe and thus were physically much better adapted to coping with the local environmental conditions than the new immigrants. Nonetheless, Neanderthals had vulnerabilities. They seem to have been specialized in their diet, depending almost exclusively on the hunting of large herbivores. Crucially, despite the proposed shorter reproductive cycle, their population size appears to have always been small, and they regularly sustained high levels of trauma and mortality.

Modern humans were probably favored because of their more flexible diet and their superior technology. While they often ate the same large mammal species as Neanderthals, early

H. sapiens also hunted the more elusive small mammals, birds, and fish. The more varied modern human diet probably explains how the two species were able to coexist in the same geographic region for as long as ten thousand years. But perhaps our ancestors' greatest advantage was demographic. People today share unique demographic features: we take even longer than the great apes and early hominins to grow, and we have extraordinarily long lives. Unlike apes, we take a short time between births: two babies can be born to the same mother less than a year apart in developed nations and within about three years in present-day hunter-gatherer societies. A chimpanzee mother will give birth only every six years or so. This human characteristic has been linked to our species' longer life span: older adults, especially women, by contributing to child care, allow women of reproductive age to have more children.[3] While we don't know whether early modern humans had shorter interbirth intervals than Neanderthals, we do see an increased survival of older adults in the fossil record for the first time with early *H. sapiens*, implying a demography similar to ours. Such a demographic feature would have supported the population increase we see in the archaeological record and would have fueled the great dispersal of our ancestors throughout the world in the last sixty thousand years. This alone would have been enough to crush the small, stable Neanderthal populations, without a single blow.

[3] Hawks, K., "Grandmothers and the Evolution of Human Longevity," *American Journal of Human Biology* 15(2003): 380–400.

And Then There Was One

Modern humans replaced Neanderthals in Europe, and probably also all other archaic humans alive at the time of the expansion of *H. sapiens* out of Africa. Even before appearing in Europe, our ancestors had reached Australia, and a few thousand years later they established themselves in the Americas. None of those continents had previously been colonized by hominins. Our generalist diet, our ability to survive in a wide range of environments, and our peculiar demography have resulted in vast population growth and expansion to the far corners of the globe in just sixty thousand years—less than a heartbeat in geological terms. Now there is nowhere else left to go. We are the sole survivors of our line. We number in the billions, and our activities have been taxing Earth's ecosystems and most probably shifting its climate. The consequences are dire for increasing numbers of organisms, but what are they for us? Will the new conditions we have created be able to sustain us, or have we inadvertently dealt ourselves the same fate that met our competitors in the past? One thing seems clear: we must once more adjust our remarkably flexible behavior to meet the unprecedented challenge of climate change and a biodiversity crisis approaching the scale of mass extinction. Perhaps there is still time to adapt.

GAVIN SCHMIDT

is a climatologist with NASA's Goddard Institute for Space Studies in New York, where he models past, present, and future climate. He received a B.A. in mathematics in 1989 from Oxford University and a Ph.D. in applied mathematics in 1994 from University College London. He was a postdoctoral fellow at McGill University, in Montreal, until 1996, when he was awarded a Climate and Global Change Postdoctoral Fellowship from the National Oceanic and Atmospheric Administration and moved to the Goddard Institute.

Schmidt was cited by Scientific American *as one of the fifty leading researchers of 2004 and was a contributing author for the 2007 Nobel Prize–winning report of the Intergovernmental Panel on Climate Change. He is a cofounder and contributing editor of RealClimate.org, which provides context and background on climate-science issues that are often missing in popular media coverage.*

WHY HASN'T SPECIALIZATION LED TO THE BALKANIZATION OF SCIENCE?

GAVIN SCHMIDT

Scientists are those who know more and more about less and less until they know everything about nothing.
—John Ziman

Reductionism works. Two thousand years of scientific progress have shown that breaking problems down into smaller and smaller bits to get at the essence of complex phenomena is a very good way to understand the natural world. However, even as scientific output has increased exponentially, concerns have been raised that growing specialization will end by making it impossible for scientists in different fields to communicate, let alone collaborate.

Yet despite dire prognostications, interdisciplinary science has been and continues to be a source of great insight. The forces driving toward ever-greater reductionism are countered by forces that foster cross-cutting research. I will try here to illuminate the scientific and social factors at work by looking at how new specialities are created and how they feed information back into the general body of science.

By the end of the eighteenth century, it had become impossible for one person to keep up with all the new scientific publications. It has been suggested that the physicist, physician, and Egyptologist Thomas Young (1773–1829) was the last person to "know everything." Another candidate is the archaeologist, mathematician, biologist, physicist, vulcanologist, and (of course) Egyptologist Athanasius Kircher (c. 1600–80). It matters little who the last universal polymath was—or if, indeed, such a person ever really existed. But with the burgeoning of scientific journals in the early nineteenth century, scientists began complaining about the divergence in research and decrying the lost opportunities for the cross-pollination of fields. Increasing specialization was blamed for almost any ill in the scientific realm, from falling attendance at meetings of the Boston branch of the American Association for the Advancement of Science in 1900 to the supposed isolation from the humanities discussed by C. P. Snow in his 1959 Rede Lecture, "The Two Cultures."

Although science has not broken down into a cacophony of mutually incomprehensible subdisciplines, there are certainly forces tending to fragment the scientific landscape. Some are natural, unavoidable, and to be applauded; others are institutional, avoidable, and to be deplored. But there are also strong incentives to break out of any rigid categorization. I will use the term *centrifugal* to describe the forces that drive increasing divergence and *centripetal* for those that work to counter it.

The scientific enterprise can be seen as a dynamic system, with the tension between its centrifugal and centripetal forces helping to move it ahead. My observations are drawn from my own experiences in earth science—specifically, the study of climate and climate change. A brief overview of this broad field will probably help.

Earth's climate system consists of its atmosphere, oceans, land, and ice: their dynamics, chemistry, and composition combine in myriad ways to produce our "climate," in the everyday sense of "average weather." Scientists study this system in many different ways: they directly observe tropical Pacific Ocean variability (El Niño events), they analyze satellite measurements of changes in sea level, they measure the gas concentrations in bubbles in eight-hundred-thousand-year-old Antarctic ice, and they develop complex computer models of atmospheric circulation, to name but a few.

The fundamental questions in each of these endeavors are similar: Why is the climate the way it is? How has it changed in the past? Is it changing now? Why and how might it change in the future? But the time scales, methods, and kinds of answers vary enormously. Although all climate scientists are nominally studying the same thing, the individual communities of modelers, meteorologists, and ice-core specialists have different notions of exactly what questions to ask and what the satisfactory answers should look like. Examining how interdisciplinary science gets done in this field is thus a good way to see how the centripetal and centrifugal forces affect science in practice.

Why Fields Move Apart

Like cholesterol, there are good centrifugal forces and bad centrifugal forces. The good ones spring from impeccable reductionist motives: the need to isolate chemical reactions occurring in ozone depletion, the need to pin down the details of sea-ice thermodynamics, the need to improve the modeling of radiative transfer in the atmosphere. Each of these is a full-time job, demanding specific and nonoverlapping expertise in laboratory techniques, field observations, or complex mathematics. There is no particular reason for such researchers to discuss these issues together; indeed, their day-to-day questions, problems, and solutions will predominantly interest only their similarly employed colleagues. The bad centrifugal forces spring from our human failings: tribalism, inertia, bureaucracy, and the understandable desire for job security.

New subfields often grow up around a new tool or technique that has produced interesting results. There is follow-on work by other scientists, and if the tool or technique shows sufficient promise, meetings will soon be organized specifically to look at those results, specialized Ph.D. programs will be offered, professorships and specialist journals will follow. As this evolution occurs, new jargon is coined and a template for success within the new subfield is forged. Examples abound: PCR (polymerase chain reaction) allowed for DNA fingerprinting, gene sequencing, and the Human Genome Project; computer hardware development allowed for mathematical models that predict the weather; satellites led to the field of remote sensing; the supercolliders at CERN (the European Organization for Nuclear Research) reinvented high-energy physics. This is all to the good.

A recent example of the birth of a new subfield in earth science is paleoceanography—the science of past changes in the ocean—covering changes in ocean circulation, chemistry, and biology over the last hundred million years or so. The innovation that gave rise to this field was the development of deep-ocean drilling, which has allowed scientists to extract archives of ancient ocean mud from almost anywhere on the ocean floor. The scientific breakthrough that made this tool into a science was the discovery in the 1960s that the time sequence of geochemical tracers in small shells in the mud mirrored the waxing and waning of the ice ages—and has done for millions of years. This result was used to cross-date and correlate data from every ocean basin and led to (among other things) the conclusion that the cycling of the ice ages results from predictable wobbles in Earth's orbit.

As might be expected, a new technical lexicon arose—orbital tuning, foraminiferal transfer functions, alkenone thermometry, calcite dissolution indexes, thorium-corrected sedimentation rates, and so on. More subtly, terms from other disciplines were appropriated and given new meanings: "high resolution" refers to sedimentation rates large enough to allow for estimates of climate change over a decade or a century, but this definition is very different from that used in physical oceanography, astronomy, or microbiology; "ventilation" in the deep sea has little to do with air currents but everything to do with the mixing of deep-water and surface-water masses.

It makes perfect sense for users of a new technique that sheds light on systems that have been studied for decades to pick up terms and concepts from existing fields. Yet since what is seen using the new technique is not necessarily the same thing as defined in another, the same concept can end up applied to a subtly different phenomenon. The Indian parable of the blind men and the elephant—who variously described the beast as a rope or wall or spear and so on—comes to mind. Though perhaps the inverse parable is more apt: just because one feels a rope, it does not mean that one has an elephant.

This is where one starts to see less-than-scientific factors driving the isolation of scientific subfields. A good example is change in temperature gradients across the tropical Pacific. In modern observations, these changes are associated with El Niño events, a dynamical shift of warm water from the West Pacific to coastal Peru. Observations of ancient changes in temperature gradients have been associated with similar events, but such "paleo–El Niños" are not necessarily recognizable to a modern oceanographer. To paraphrase Winston Churchill, specialities can become separated by a common language.

An additional social factor is that with the accumulation of scientific results comes a specific subculture and set of templates for how to succeed in the new field. The older, pioneering generation wants to employ and promote successors in its own mold; the younger generation wants to emulate its mentors. In both cases, a pattern emerges that should be followed in order to win the acclaim of one's peers within the field. In paleoceanography, that pattern was set by the data-generating geochemists who started the field forty years ago. Their reputation was based on their solid analytical skills and tied to data from a few locations in the ocean, to which they had exclusive, or near exclusive, access. Ocean drill cores necessarily provide a time sequence at a specific point, and they can be difficult to compare in detail with other cores, because of the uncertainty in assigning an age to any particular depth. Developing the "age model" for a particular core is often the most challenging part of putting new data in context. Thus the vast bulk of papers in the field discuss transient changes at specific locations rather than making the much more difficult assessment of the geographical patterns of change at any one time. The path for aspiring young paleoceanographers is clear: you attach yourself to a consortium of ocean drillers and invest in the equipment that will let you make your own high-precision measurements. This is usually some kind of mass spectrometer, which costs around one million dollars and is often part of the negotiated package for new faculty hires. But with this equipment come responsibilities: to fund technicians and bring in interesting samples, in order to justify the use of laboratory space.

Thus a young researcher is locked into a mode of working based on the dominant pattern within the field. It can be difficult to make a career doing a different kind of science. In particular, the very verticality of ocean cores—a function of how samples are gathered and analyzed—makes it problematic to study changes in a horizontal way, since that information is widely distributed. This discourages interdisciplinary efforts, because the broader field of climatology works very much on such "horizontal" questions as the geographical patterns of climate change.

To summarize, once a subfield becomes established, its necessary and useful specialization puts up a barrier to interdisciplinary work: jargon excludes most of those outside the subfield; subtle miscommunications occur when terms are borrowed and redefined; specialized journals, conferences, departments, and investments in equipment propagate and deepen the use of the new methodology. All this can create an isolated subculture whose conversations are increasingly estranged from those in the wider field. The individuals involved are doing nothing obviously wrong, but, like "the tragedy of the commons" in economics, the collective impact of rational individual decisions can lead to undesirable outcomes for the field as a whole. Which in turn can lead to a lack of outside appreciation of the field's successes and reduced access to research funds — and internally to a sense of frustration that the message "isn't getting out."

The Path to Synthesis

To see how the centripetal forces of scientific integration work, it's helpful to look at a discipline from the perspective of scientists in related fields. How do they find out about new results in fields related to theirs? How do they perceive those results? And what motivates them to get involved at a more fundamental level?

There are three main channels through which information about new scientific findings percolates: popularization, "ambassadors," and publication in the technical literature. With regard to the first of these, it might surprise you to learn that (for instance) many more climate scientists will have read James Gleick's bestseller *Chaos* than Edward N. Lorenz's original papers on the subject. In my own attempts at popularization, I am surprised by how often colleagues in the same general field tell me they found my nontechnical descriptions useful for clearing up important points of detail.

The second channel is made up of those scientists with a technical background in one discipline who spend much of their time talking to a wider group of scientists and who can sometimes embed themselves within another field. They are strong advocates for collaboration, since they can often translate effectively between fields; however, they are few and far between.

With respect to the technical literature, scientists, like the lay public, find it is generally impossible to read (let alone judge the worth of) papers at the cutting edge in a different field. Instead, the papers they notice tend to fall into three main categories. First, there are the studies that deal with something really big and get attention simply because of their intrinsic interest—for instance, the 1980 paper by Luis and Walter Alvarez about the asteroid impact at the end of the Cretaceous period, or Kate Spence's 2000 study using the precession of Earth's orbit and a rule of thumb for judging true north in Ancient Egypt to independently date the building of the pyramids. However, while iconic, these papers do not necessarily invite collaboration.

The second kind of paper that gets noticed is a technical paper that initially appears to be more broadly relevant—for instance, those reporting on past hurricane landfalls in Massachusetts, or El Niño events six thousand years ago, or the possible human role in the extinction of large mammals at the end of the last ice age. Each of these studies could well be relevant to someone studying hurricanes, El Niño, or extinctions in the modern world, though the direct relevance is often small. The location studied is unlikely to be representative of a larger area; the time period or the resolution is likely to be too limited; the circumstances in which data were taken may be too complicated or unclear. Sometimes the aforementioned subtle differences in definitions can cause communication problems. These kinds of studies lead to successful interaction only occasionally—for instance, when there is a real overlap, such as when a paleo-record lines up with more standard historical information.

The third and most important group of high-profile papers are synthesis results—studies that boil down the essentials of what has been learned from many individual studies and put them together. (Unsurprisingly, the best ambassadors also tend to be good synthesizers.) These syntheses can be metastudies, like those seen in epidemiology, or a "horizontal" reconstruction of a particular past climate that correlates many disparate records. Often the synthesis involves numerical models, such as the assimilation of different kinds of weather data by models to produce consistent estimates for the state of the atmosphere. Development of large-scale models is itself a primary example of synthesis, bringing together computational, atmospheric, oceanic, and cryospheric experts.

A good synthesis paper often provides a convenient jumping-off point for a lot of interdisciplinary research. This is certainly true of paleoclimate syntheses—for instance, through the mapping of temperatures and ice sheets at the peak of the last ice age, the linking of ice-core records from Greenland and Antarctica, and the reconstruction of patterns of climate variability in recent centuries. These papers have led to extensive modeling efforts to understand those patterns and have encouraged complementary avenues of research. Such papers can often dramatically increase the profile and relevance of a field in the wider community—key factors in increasing the resources available for further research and appreciation of the discipline's relevance.

The search for centripetal forces in science is roughly equivalent to asking why and how synthesis happens. The need for synthesis arises as the number of individual data points or physical processes becomes large enough to make them difficult to keep track of. Synthesis, basically, is concerned with cutting through the clutter. Individual scientists differ tremendously in their attitude toward integrative analysis (after all, one person's noise is another person's signal), and that difference is reflected in who chooses to pursue synthesis projects.

Such projects can be built either from the bottom up (that is, within the field) or from the top down (at the behest of outsiders or institutional interests). The estimates of the climate of the last ice age were derived bottom up, from a consortium of paleoclimate researchers. Conversely, in oceanography, the highly cited *World Ocean Atlas* of temperature and salt content was compiled from millions of individual data points by scientists at the the National Oceanographic Data Center rather than by any of the data-gathering physical oceanographers themselves. Outside pressure is, in fact, a very common driver. It can come from modelers who need integrated data for comparison with their simulations or from assessment processes such as those run by the Intergovernmental Panel on Climate Change or the National Research Council.

Despite the wide influence of such syntheses and the impressive amount of follow-on science they generate, common to almost all of them is a critical, sometimes hostile, response from within the field. Frequently it is asserted (not unjustifiably) that the synthesis simply cannot be acceptably done, because there are too many unknowns and too many confounding factors. There is grumbling that the synthesizers are enjoying the fruits of the labor of the data gatherers without having done any of the hard work themselves. Outside requests for synthesis tend to be met with complaints that subtleties will be neglected and lead to misunderstandings, particularly among the public.

Understandably, given their sometimes difficult birth, syntheses are subject to close scrutiny, mainly to good scientific effect. For instance, the first tropics-wide estimates of ice-age ocean cooling were shown by modelers to be inconsistent with land-based temperature estimates, leading to a decades-long reassessment of the methodologies used and eventually to a new bottom-up synthesis.

Ironically, the works that form the outside perception of a field are often resented internally, and support for further syntheses can often be hard to marshal. This is where the centripetal forces most clearly face the centrifugal forces, both good and bad. Is it worthwhile to devote time to synthesis or not? Without it, progress in the field can still be impressive and new insights will likely continue, but outside appreciation may lag, leading to growing frustration internally. This very frustration can become a centripetal force, prompting scientists to collaborate to share what they have found. The predominant mechanism of that sharing is synthesis, and the easier it is to do, the more interdisciplinary science becomes.

Can this process be sped up or encouraged? I think it can. Recognition of the centrifugal forces that hinder synthesis and communication can lead to specific mechanisms to counter them. For instance, funders can set aside money specifically for synthesis, so that such projects don't compete with the funding of technical research; data can be made more accessible, with the goal of synthesis in mind; the benefits to the field of greater outside interest in its work can be emphasized.

Fundamentally, the drivers of interdisciplinary science lie in our desire to explain what we see and in recognizing that the answers we seek are not tied to individual scientific disciplines or specific tools and methodologies. Those are human constructs, and they are simply no match for the forces of nature.